DATE			

SHOEMAKER BY LEVY

.

SHOEMAKER BY LEVY

The Man

Who Made

an Impact

David H. Levy

PRINCETON UNIVERSITY PRESS

PRINCETON AND OXFORD

Library of Congress Cataloging-in-Publication Data

Levy, David H., 1948–
Shoemaker by Levy: The man who made an impact /
David H. Levy.
p.   cm.
Includes bibliographical references and index.
ISBN 0-691-00225-8 (alk. paper)
1. Shoemaker, Eugene Merle, 1928–1997 2. Astrogeologists—
United States—Biography. I. Title.
QB454.2.S48   L48   2000
559.9′092—dc21
[B]   00-038523

This book has been composed in Sabon and Futura Typefaces

The paper used in this publication meets the minimum requirements
of ANSI/NISO Z39.48-1992 (R1997) (*Permanence of Paper*)

www.pup.princeton.edu

Printed in the United States of America

10   9   8   7   6   5   4   3   2

For Wendee

MY COMET SHINING IN THE NIGHT

WE SHARE AN ORBIT

CONTENTS

This was the noblest Roman of them all: . . .

His life was gentle, and the elements

So mix'd in him that Nature might stand up

And say to all the world, "This was a man!"

—SHAKESPEARE, Julius Caesar, *1599*

WERE IT NOT for the fate of a comet thirteen thousand years ago, the Shoemakers and I might not have become a team. But on the afternoon of September 16, 1968, comet finding was far in our future, the United States was preparing to send men to the Moon, and I was sitting in the office of George Stevens, head of Acadia University's geology department. As a twenty-year-old amateur astronomer who hoped to find a career in science, I needed advice.

There's a new field that connects astronomy to geology, George explained; it's called astrogeology, and its heart and soul is a man named Eugene Merle Shoemaker, at the U.S. Geological Survey (USGS) in Flagstaff, Arizona. That's the first I had heard of Shoemaker and his life's work, to show that impacts from comets and asteroids had substantially affected the course of life on Earth. Twenty years later, I finally met him at the Asteroids II conference in Tucson, Arizona. The meeting focused on the solar system's minor rocky bodies and their effects on the planets. I stood at one end of the coffee room, he at the other, but with a laugh that could be heard for blocks, Gene could not be missed in such a crowd. We made eye contact, and then he pointed toward me as if we were old friends—I turned around to make sure there wasn't someone else he wanted. "David!" he called in his forceful way, "You're the man I want to meet!"

Actually, *he* was the man *I* wanted to meet. If he was already well known in the late sixties, by the mid 1980s he was a towering figure in science. Although he accomplished many things in many areas, Gene's life and career were tightly focused on the single subject of cosmic impacts. Early in his career, Gene became interested in collisions after noticing the similarity between atomic blast craters he had seen and the large crater near Winslow, Arizona. He confirmed that this crater was formed by an impact from space. Throughout his life, Gene continued to explore craters on Earth, from Australia to Saudi Arabia. Earth was not his only target, and in 1948 Gene turned to the Moon as an object of comet and asteroid bombardment. With more than three hundred craters visible even through a backyard telescope, the Moon was a fertile place to study the results of impacts. With men about to head to the Moon, Gene was an ideal scientist to train the astronauts in what they might find there. After Apollo, he joined the Voyager imaging team as it studied each new Voyager spacecraft image of the crater-strewn satellites of Jupiter, Saturn, Uranus, and Neptune. In the 1970s Gene turned the tables on his own studies, switching from the craters to the objects making the craters. He wanted to determine the flux of comets and asteroids in Earth-crossing orbits, objects that could someday end their lives, and possibly ours, in fiery collisions with Earth.

I knew what he had accomplished, but I didn't know the man. Within a few seconds he had introduced me to Carolyn, his wife and career partner. He made me feel as though we had been good friends all our lives. He quickly got to his concern, that the eighteen-inch telescope on Palomar Mountain north of San Diego, California, the telescope the couple had been using for their comet and asteroid search, might not be reliable. They were looking for a backup telescope, and they thought that with my own experience in comet hunting and observing, using telescopes here, I could help.

As if to cement the idea, two weeks later I discovered that ancient comet, now called Comet Levy 1988e. On hearing of my find, Gene and Carolyn decided to add it to their observing program at Palomar. On that night they finished their regular program after dawn had begun. They moved the telescope over to the east and took a brief exposure of the field that contained the new comet. Despite the brightening dawn sky they got a good image and submitted the first accurate positions of my comet.

The next month Gene and Carolyn were accompanied by their colleague Henry Holt, and again they included my comet on their list of fields of sky to photograph. They set up their list graphically on a sheet of paper on which dozens of nickel-sized circles outline the observing fields around the sky. But the comet was north and east of where they usually photograph, so Gene placed an extra circle at the top of his "nickel diagram."

On the morning of May 13, 1988, the telescope was pointed at the position indicated by the extra circle. The following evening, Carolyn placed the Comet Levy films on her stereomicroscope and quickly found a comet. Gene quickly had a look, but wondered why the comet was so far from the field's center. Gene then went back upstairs to continue setting up for the night's work, while Carolyn wondered why the comet was not at the center of the field as she had planned it. After a few minutes she called upstairs again: "Gene, I cannot identify any stars on this film. I don't know what field this is!" Gene was already guiding on a star at the telescope. "Henry"—he quickly handed over the guide paddle—"take over." After he rushed downstairs it took a him a few minutes to identify the mistake: they had taken the field represented by the position of the circle on the nickel diagram, not the position where Comet Levy was—a substantial difference of seventeen degrees. Satisfied that he had solved the riddle, Gene was about to go upstairs again when Carolyn asked the question, "But if this comet isn't David's, then whose is it?"

It turned out that they had discovered a new comet, Shoemaker-Holt 1988g. After enough positions of the new comet allowed a calculation of the interloper's orbit, Conrad Bardwell of the Central Bureau for Astronomical Telegrams, noticed an extraordinary similarity of its orbit to that of Comet Levy. The two orbits were almost identical in every respect except that Comet Shoemaker-Holt arrived at perihelion some three months after Comet Levy's closest approach to the Sun. This was a singular instance of a pair of related long-period comets being discovered independently. The two comets were one some twelve thousand years ago, then split apart.[1] It was also a singular message from space that Gene, Carolyn, and I belonged together.

We observed together for the first time a few months later at a sixteen-inch telescope on Mount Bigelow near Tucson, although that telescope didn't work out as we had hoped. In July 1989 I began observing regularly with the Shoemakers at Palomar. For perhaps six months of the year, I'd meet them at the observatory and we'd spend a week using the telescope in an intensive search for asteroids and comets.

It was during these many observing runs that I got to know Gene and Carolyn. "Camping or observing," Carolyn loved to say, "are the best ways to get to know a person really well." On that criterion, we got as close as one might expect, going through happy and sad times together. We got to know each other this way through increments. Each picture we took of the night sky was a time exposure that typically lasted eight minutes. During at least part of that time one of us would be loading a new film while another would guide the telescope as it recorded the star images. That process would leave maybe four or five minutes during which we'd just have time to talk. I might ask Gene a question about impacts, or a past experience somewhere, and he would answer, always being sure that his words ended before the exposure did. It took me a while to realize that even during our busiest observing sessions, I left knowing Gene and Carolyn even better than before.

The third week of July 1994 was the most intense that we had ever experienced. Sixteen months earlier we had discovered Comet Shoemaker-Levy 9, and during that magic week we, and the world, watched as the object's twenty-one fragments pummeled Jupiter. We observed the results through visual telescopes and, via computers, through the Hubble Space Telescope; we were interviewed for almost every network and newspaper; we made ABC's Persons of the Week, and we met President Clinton and Vice President Gore. By the time our observing program ended at the end of 1994, our friendship had gone well beyond a close one.

CITIZEN SHOEMAKER

The purpose of a biography, Orson Welles intimated in *Citizen Kane*, is to show what a person is like. I have tried to accomplish this goal in an objective way, but since the man you are about to meet was very special to me, this book is by no means a journalistic biography. (I also had a much easier task than the reporter in *Citizen Kane* did, since Gene was a far more decent person.) I believe that by drawing attention to the role of cosmic impacts, Gene Shoemaker fundamentally changed the way we look at the planet we live on, and the story I have to tell will show how he did this. As youthful experiences and indiscretions shape all of us, the book's early chapters show how Gene's school and college days set him on the track to developing into the scientist he was destined to become. His courtship and marriage are important times, for Carolyn was far more than his wife; she was his scientific soulmate, without whom he would not have accomplished much of his later work. Gene's years with the U.S. Geological Survey pointed his thoughts toward the geology of the Moon and other worlds.

Gene's hallmark work showed that Meteor Crater, the large hole in the ground near Winslow, Arizona, was the result of impact. But this early accomplishment merely crystallized the direction his own career was taking. He did much more than hitch a ride on the unmanned Ranger and Surveyor lunar probes, and on Project Apollo's manned landing—by his own leadership and insight he helped direct where those craft would land and what research they and their crews would perform. With the end of Apollo in 1972, Gene's journey forked to new but related directions. In Australia he studied the impact craters that resulted from comet and asteroid impacts on Earth, and on Palomar Mountain he searched for the asteroids and comets that might be responsible for future impacts. This effort was the culmination of his lifelong interest in the question of impacts on Earth.

Through this book, I hope you will get to know what Citizen Shoemaker was like: his story, his accomplishments, his dreams and failures, his moods—everything that makes up the special man we remember as Gene Shoemaker. Through observing nights and other experiences, I was a colleague of Gene's and Carolyn's for a long time. I was drawn to them. Once, early in our friendship, I asked Gene about their next observing session. "We're going after those Trojan asteroids near Jupiter!" he exclaimed, proceeding to wave his arms about as he explained how these distant asteroids orbit the Sun either behind or ahead of Jupiter, and that they've become a special interest. "We want to be like Johnny at the rat hole when these asteroids first appear in the morning sky!" He often used that expression, and I think it was appropriate for a man who wanted to be the first in line to go to the Moon, the first in line to show that Meteor Crater was the result of an impact, the first in line to discover comets with Carolyn. I loved his total passion and commitment to what he and Carolyn were trying to accomplish. Late one night as I sat on a stool closely watching the motions of the "guide star" the telescope was following, Carolyn was watching the clock to tell me when to end the exposure. We were feeling good since we had just discovered a new asteroid (to be called 1989 VA) moving rapidly through the Pleiades cluster. Then Gene opened the door from the darkroom downstairs and walked up the flight of steps with a film holder all prepared for the next exposure. As he walked past Carolyn, he smiled, and they held hands and embraced. It was so wonderful be be a part of this adventure that I gave up a paying job to continue to join the week-long observing sessions each month. Gene had so much more than simple energy and enthusiasm; his personal magnetism could envelop a situation. Again and again, his colleagues, friends, and family have told me how that quality of his got them to do so much more than they ever thought they were capable of accomplishing. This was Gene Shoemaker: a man with the passion to make his science fun, special, and full of love.

1. Brian Marsden, "The Comet Pair 1988e and 1988g" (1988).

Writing a biography of a person's life is an intensive and personal task; during the time set aside for it the writer needs to step into the subject's personality, to learn what it was like being the subject. I would never have begun this project had I not had experience with two earlier subjects, so I would like to begin these acknowledgments by thanking them—Clyde Tombaugh, discoverer of the planet Pluto,[1] and Bart Bok, the scientist of the Milky Way—for helping frame my thoughts as to what a biography should accomplish.[2]

Art Boehm provided a careful reading of the original draft of the manuscript for this biography. His suggestions improved the text immeasurably. I appreciate the insights provided by George Stevens at Acadia University on the geological background chapters 5 and 6, and the strong suggestion from David and Raili Taylor in July 1997 that I begin this book. I thank the many people who provided insights into Gene's life and personality, including J. Kelly Beatty, Fred Bortz, Donald Brownlee, Steve Dwornik, Donald and Shirley Elston, Ronald Greeley, Alan Harris, Eleanor Helin, Susan Kieffer, Joe Kirschvink, Candace Kohl, Larry Lebofsky, Peter and Margaret Marsh, Clifford Matthews, William Muehlberger, Jean Mueller, Carolyn Porco, Justin and Renate Rennilson, Frank Rock, Harrison Schmitt, Bob Sharp, Leon Silver, Gordon and Jodi Swann, and Jeff Wynn. I also thank Bevan French for his gracious permission to let me use his poem, *The Man Passing By on My Way to the Moon*. Gene's family was very helpful—especially his children, Christy, Pat, and Linda; his son-in-law, Phred Salazar; his daughter-in-law, Paula Kempchinsky; his granddaughter Stefani; and his grandson Sean. Gene's sister, Maxine Heath, and her husband, Jim, and Carolyn's brother, Richard Spellmann, also were most helpful. I am also thankful for the help and support I received from Jack

Repcheck and Kristin Gager at Princeton University Press, and from Clark Chapman and Paul Chodas, who reviewed the manuscript for the Press.

Without the help and support of two people this book would not have happened. One is Carolyn Shoemaker, who supported the project years before it began and who read the manuscript thoroughly. The other is my wife Wendee Wallach-Levy. Besides her careful manuscript reading, she kept my writing schedule intact, checked the final proof, and helped with the index. From the bottom of my heart, thanks.

1. Levy, David H., *Clyde Tombaugh: Discoverer of Planet Pluto* (Tuscon: University of Arizona Press, 1991).

2. Ibid. *The Man Who Sold the Milky Way: A Biography of Bart Bok* (Tucson: University of Arizona Press, 1993).

SHOEMAKER BY LEVY

CHAPTER 1

Of Bonding and Discovery: 1993

Strange, is it not? that of the myriads who

Before us pass'd the door of Darkness through,

Not one returns to tell us of the Road,

Which to discover we must travel too.

—Fitzgerald, Rubaiyat of Omar Khayyam

Do you three ever have a comet!" said Jim Scotti from the control room of his telescope. Shielded from the cold night atop a mountain in southern Arizona, Jim was looking at his monitor screen and beholding the strangest sight he had ever seen in the night sky—a train of several comets, each one coupled to the next by a thick layer of dust. Gene and Carolyn Shoemaker and I were in a room beneath our telescope, four hundred miles to the west on Palomar Mountain, also huddled against the wind and snow outdoors. We would not be opening the telescope this night, but thanks to Jim's confirmation, our cometary prey was already in the net.

Two nights earlier, the evening of March 23, 1993, Gene; his wife, Carolyn; and I were at Palomar Mountain Observatory, working on his Palomar Asteroid and Comet Survey. We were taking photographs of the sky—as much of it as possible each month—in search of comets and asteroids. Gene's goal was to determine the rate of impacts in our corner of the solar system.

Gene had a dream. With all the craters, and crater-making objects he had seen, he wanted to cap his career by actually witnessing an impact, perhaps on some remote desert area in Australia, and then rush out and map the result. He talked about that dream often, and even might have been thinking of it on that evening of March 23, when a row of clouds put a stop to our observing. We walked outside, together with Philippe Bendjoya, an astronomer from France who was checking out our operation that month.

"I think we should keep on observing," I said, conscious that in an El Niño year, the onset of clouds likely meant that we would not get to use the telescope for several days.

Gene's reply had a financial twist to it. "There's David, ever the optimist!" he began. "But really, we spend some four dollars each time we slap a piece of film into that telescope. In the course of a year that might amount to some eight thousand dollars just on film. David, we're just going to have to wait this one out."

As we stood there silently, gazing up at what stars could be seen through the gathering clouds, I remembered the rude surprise we had received the evening before: we had to stop observing because our film had gone bad. Someone had opened our box of carefully prepared emulsion, exposing its contents to light. The films had been stored one on top of the next, so the uppermost films, which were completely ruined, served somewhat to protect the lower ones. We limped through the rest of that night with these damaged films—their centers were usable, but their edges were struck with light.

Our second night began on a better note. Armed with a fresh supply of film, we began work and kept on until those clouds rolled in. And here we were, waiting out what appeared to be the leading edge of another storm. I thought about last night's damaged films, and of Gene's financial argument, and had an idea.

"Gene, do we have any damaged films left over from last night?"

"Yes, I saved them for focus tests."

"Those films don't cost us anything," I continued. "If we use them, we have nothing to waste but our time."

We looked at the sky, then at each other. "Let's do it!" Gene bellowed. We rushed up to the dome. Gene gave me the filmholder, which I locked into place in the belly of the eighteen-inch Schmidt camera. He read me two coordinates, the right ascension and the declination—the sky's extension of longitude and latitude—and I positioned the telescope. I looked through the eyepiece to locate the star on which I would guide the eight-minute exposure and was really surprised to see a bright glow. I looked hard and found the guide star, barely visible amidst the glow. I looked up and saw the problem at once: Jupiter, the solar system's largest planet, was very close to the center of where I had pointed the telescope.

I suggested to Gene that because the partly clouded sky was so bright from Jupiter's glare, maybe we should move to another field of sky. But since this was the next field on our program, we decided to stick with it. "If you can't see the guide star at all," he said, "then we'll go somewhere else."

"No," I answered, "the star is faint, but I can follow it. We can stay here. I'm ready to go!"

Gene then waited until the clock's second hand was about to reach a quarter minute, and then began to count: "Five, four, three, two, one, Open!"

How to Learn a Lot in Four Minutes

Pulling on a lever, I heard a soft squeaking sound as the telescope's two metal shutters swung open. The exposure was in progress. Gene went downstairs to load another sheet of six-inch diameter circular film into its holder, then returned. For about three minutes, Gene could relax as I guided the exposure. We chatted casually.

These three-or four-minute chats were extremely useful times for us. While I can't recall what we talked about during this particular exposure, these times were golden opportunities to get caught up on news, to express our opinions about various subjects, or just to enjoy the night. Conversation was important during this time, especially as the night wore on and we grew tired. We also had a cassette recorder that played classical, jazz, country, or Australian music.

Sometimes—especially late at night—the conversation would be just plain fun. When Carolyn and I observed together we got downright nonsensical—the humor kept us going and made the long nights pass quickly. On our very first observing run together, Carolyn and I were at the telescope while Gene was in Pasadena as part of the Voyager 2 imaging team looking at images coming in from the spacecraft as it swung by Neptune's moon Triton. Carolyn had prepared two boxes of films for use that night—using a film cutter to shape the films into six-inch circles, she then connected the boxes to a supply of "forming gas" of 92 percent nitrogen and 8 percent hydrogen, and placed them into an oven. In six hours, the films were hypersensitized—which meant that with our Kodak 4415 Technical Pan film, the hypersensitizing process increased the sensitivity of films to light by a factor of nine. Gene called the process "baking the cookies."

On that second night together, our program was proceeding faster than expected. Thinking that our supply of films was running low, Carolyn mused that we might actually have to stop before dawn if we ran out of hypersensitized film. So as Carolyn guided an exposure, I went downstairs, replaced the film, and retraced the twelve steps to the observatory dome. "Carolyn," I said plaintively, "I checked. We're completely out of film."

"I see," she replied.

"The film you have is the last one."

"Hmmm."

"But I think I found a way to use more film."

"What's that?"

"Two minutes to go in this exposure, Carolyn. Well, you know I've been using hypered 35mm film in my own camera . . ."

"Yes?"

"Well, I brought some. What I did was, I took four strips of the film, crunched it into the filmholder, and like that!"

Carolyn looked really puzzled. "How do I tell this nice man," she later admitted thinking, "who has come all this way to be a part of the program, that his idea is just plain dumb?"

Instead she said, "But David, when we develop these strips of film, how shall we know which one goes where?"

"Forty-five seconds, Carolyn," I noted the time remaining on the current exposure. "Well, we can put them together with Scotch tape and then I'd compare them with a nice star Atlas and we could figure it out." Really concerned now, Carolyn looked up from her eyepiece and noticed my wide grin.

"David! I think I've just been had!"

"Twenty seconds!"

We laughed about that one for years. It was the first of many moments of genuine laughter between us that helped make these moments so special. The best jokes came late at night. On another night I rushed up the twelve stairs to the dome to present Gene with a new film, only to realize that there was still a few minutes left of the exposure. "You're in a hurry!" Gene said, his nimble fingers pressing the buttons that guided the telescope. Since this particular exposure was in the eastern half of the sky, where the telescope tended to wander off its guide star, Gene was pressing the east and west guiding buttons every few seconds. With each press the telescope made a *sssssst* sound as it raced to catch up with the Earth's rotation, and a lower pitched *bzzzt*! if it needed to slow down.

"Well, maybe the night will go faster if I hurry," I admitted, quite out of breath.

"Well, I don't know. Did you hear about the old bull and the young bull?" *Sssst, sssst.*

I admitted that I hadn't.

"Well," *Bzzzzt*! "Damn! Wrong way." *Sssssssssssssssssst.* "Two bulls were standing at the top of a hill, an old bull and a young one." *Ssssssst.* "They were watching a herd of cows at the bottom of the hill, and the young bull was strutting around." *Ssssssst. Bzzzzt! Bzzzzzzt*! "'I wanna charge down the hill and make love to the first cow I run into!'" *Bzzzzt! Bzzzzt*!

"The old bull looked out over the herd. 'Well, you can do that. I will walk downhill slowly and sagely.' " *Ssssssssssst!* " 'And then I'll get the whole damn herd!' Damn!" *Sssssssssssssssssssssst!*

"One minute, Gene!"

Usually our conversations were more serious, like the time I asked Gene what he thought of Frank Drake's famous equation on how common life is in our galaxy. The equation reads

$$N = N_* \, f_\mathrm{p} \, n_\mathrm{e} \, f_\mathrm{l} \, f_\mathrm{i} \, f_\mathrm{c} \, f_\mathrm{L}$$

where N is the number of intelligent civilizations in our galaxy capable of communicating with us. N is based on numbers and fractions that express the number of stars in our galaxy, the fraction of those stars with planets, the number of those planets that are Earth-like, the fraction where life took hold, the fraction where life evolved to intelligence, the fraction where life evolved to technological communication, and finally, the fraction of time a civilization is likely to last. The traditional Ns vary widely; optimists suggest several hundred thousand; pessimists say one.

I found that while Gene did think that N is more than one, he feared that it is very low. He explained that the equation leaves out one important factor that might reduce the value of N. A solar system must have at least one Jupiter-sized planet. "Jupiter works like a vacuum cleaner," he said. "The planet's gravity is so strong that it rids the solar system of billions of comets." These comets, he continued, are affected by Jupiter's gravity and are flung out of the solar system, or, in rarer cases, actually collide with the giant planet. Without a Jupiter in the system, comets would continue to rain down on Earth-sized planets, just as they did in our system when it was young. Without Jupiter, our Earth would still be a target for ten kilometer-diameter comet strikes at the rate of one per century instead of the present comfortable rate of one each hundred million years.

"Thirty seconds!" Gene interrupted his train of thought to let me know to get ready to close the shutter. With time running out for the exposure, I asked Gene how serious this consideration was. "Pretty serious," he answered. I think it brings the value of N pretty close to one. I don't think we are alone, but I do think that the development of advanced life forms is pretty rare in our galaxy.

In this way, I learned an awful lot of planetary geology in three- or four-minute segments.

Out of little stories like these were wonderful memories built. I learned more about geology, impacts, and the character of Gene and Carolyn during these short segments—all while the telescope patiently gathering photons of stars, galaxies, asteroids, and comets—than at any other time during our relationship. And Gene, Carolyn, and I never let the stories overstretch the exposures.

Whatever jokes, stories, educational anecdotes, or other conversation was happening on this night of March 23, 1993, the time for the Jupiter exposure finally ran out as Gene called out "five, four, three, two, one, *close*!"

THE NIGHT GROANS ON

The sky clouded over again as we struggled through the next two images. Finally we had to stop. I went outdoors several times to check on the sky. We worried that if we did not begin the "bottom half of the inning" soon, we might have to scrap the set. The way our films work for comet discovery is that they detect the motion of a comet or an asteroid; it is in different places from one film to the next. When Carolyn stereoscopically studies a pair of films that are identical in every regard except that they were taken some forty-five minutes apart, she can see a moving object appear to "float" on top of the background of stars. What an elegant way to search for new worlds!

In order to see these new worlds, we needed two films of every area, and ideally the films should not be spaced shorter than forty minutes apart, or longer than about an hour. As we waited for the clouds, an hour passed from the start of the first exposure, then ninety minutes. Despairing, I went outside one more time. There appeared to be a small hole in the clouds, moving eastward. With a little luck, I thought, it would pass the Jupiter area and stay clear long enough for me to just get in the second exposure. Gene and I ran up the stairs, loaded the film, and reset the telescope. For a while I couldn't see the guide star at all, but after a few more minutes the break in the clouds reached the Jupiter region and the guide star just slightly appeared. I began the exposure and kept going for eight minutes. The clouds closed over again just before the end of the exposure. We struggled through the rest of that night, and on the following night we took a few more exposures before the heavier clouds moved in, shutting off our observing run for the next few nights.

A Unique Discovery

Even though we were not accomplishing much because of the weather, we were thoroughly enjoying each other's company this run. Philippe Bendjoya, one of the most delightful people we've ever shared the dome with, was interested in watching what we were doing, yet each cloudy evening he would assist me with my faltering French. Our late night dialogues covered politics, religion, even astronomy—and were all *en Français*. So by the afternoon of March 25, we were a quiet but satisfied group, saddened only because the weather was keeping us from observing. Carolyn even looked up from her scanning, saying plaintively, "You know, I used to be a person who found comets."

"Used to be?" Gene asked.

On a few occasions, when she said those words, Carolyn would find a comet. She completed her scanning of the first night's films with no luck. It was time for Gene and me to head over to the nearby sixty-inch telescope dome, which housed the oven for hypersensitizing. By now the storm was raging, with high winds and light snow. We knew that there would be no observing that night, but we were going through the motions of hypering the film and preparing the night's observing plan, just in case.

We returned to our eighteen-inch telescope dome. With observing unlikely that night, 4 P.M. was the slow part of the afternoon. I continued working on my cloudy-day activity of writing a book called *The Quest for Comets*. Gene was catching up on his reading of a paper being prepared for publication. Carolyn was scanning the Jupiter field. Somewhat tired from the previous night's observing, Philippe was napping in the car.

Suddenly Carolyn stopped the steady movement of her stereomicroscope. She had just skidded past a fuzzy something that looked a bit like the fuzzy appearance of a distant galaxy. Was it floating? She backed up, increased the magnification, sat straight in her chair, and peered intently into the eyepieces. After a few seconds she looked at us. "I don't know what I've got, but it looks like a squashed comet."

As Gene walked over to look, Carolyn smiled at me. Was she being humorous? After all, she had just defined herself as a person who used to find comets! "No," she said, "there's something strange out there."

Gene studied the images carefully, then looked at me with the most bewildered look I've ever seen on the man. Now it was my turn. I saw two discrete images of what really did look like a comet someone had stepped on. At the center was a bar of fuzzy, cometary light, appearing dark of course, since we were using the original negative films. Stretching to the north of the bar were clearly defined cometary tails. "It's got to be a comet," I called out, "look at those tails!" Then I noticed the pencil-thin lines extending along either side of the bar.

Sensing the excitement, Philippe appeared. We told him what had happened and he, too, looked at this unique object. I prepared an e-mail message about the comet; then Gene and I logged into the computer service of the International Astronomical Union's Central Bureau for Astronomical Telegrams. It was time to report our discovery to the center's director, Brian Marsden. After our initial message was cut off in midtransmission, we wrote this letter:

> We got cut off on our last message to you (the one we logged directly to your computer service) so we are resending with more details.
>
> The strange comet is located as follows:
>
> 1993 03 24.35503 12 26.7 (2000.0) − 04 04 M = 14
>
> [The numbers refer to the year, month, and date to five decimal places; the right ascension and declination in coordinates standardized to the year 2000; and the brightness of the comet at magnitude 14.]
>
> The motion is west-northwest (not southeast as in the previous message) at about 7 arcminutes per day. The image is most unusual in that it appears as a dense, linear bar very close to 1 arcminute long, oriented roughly east-west. No central condensation [a thickening of the comet toward its center] is observable in either of the two images. A fainter, wispy "tail" extends north of the bar and to

the west. Either we have captured a most unusual eruption on the comet or we are looking at a dense tail edge-on.

Right now we are sitting in the middle of a cloud with no hope of observing tonight, and we had very poor observing last night. Observers are Eugene and Carolyn Shoemaker, David Levy, and Philippe Bendjoya."[1]

After we sent this message, we left the observatory to head for the house we rented for the observing runs. During the drive back Gene suggested a possible scenario for the comet's appearance. The comet is close to Jupiter in the sky. Suppose the comet was physically near it in space, close enough to have been tidally disrupted by the planet? Maybe the comet's elongated appearance indicated that it had split apart. Like many of Gene's hunches, this one turned out to be spectacularly correct.

CONFIRMING THE DISCOVERY

After our dinner break we returned to the observatory. The sky was hopelessly cloudy, so there was no chance of rephotographing the new comet ourselves. I called Jim Scotti, a close friend who was that night observing with the Spacewatch camera on Kitt Peak, west of Tucson, as part of a search for asteroids and comets, and told him that we had an unusual comet not far from Jupiter. When I gave him the two positions, he quickly realized that the comet was also moving in the same direction and velocity as Jupiter. He suggested that our "comet" was nothing more than a reflection of Jupiter's light in the telescope! He did agree, however, to try to get an image for us.

Concerned, Gene quickly took a straight-edge and extended the line that the comet drew in space toward Jupiter. The line passed just south of the planet. If it had been a reflection, Gene reasoned, the line should have gone straight to the big overexposed splotch that was Jupiter. In the meantime, Gene and I went over to the dome of the forty-eight-inch Schmidt, which housed the device we used to measure accurate positions of our comet. Jean Mueller, our friend who works the forty-eight-inch telescope for the Second Palomar Sky Survey, greeted us there and helped us measure the comet's discovery images. About two hours later we returned to the eighteen-inch.

Because the night was stormy, we brought our music downstairs; it was playing Beethoven's first symphony. It was time to call Jim. He picked up the phone and began to clear his throat. "Jim," I said, "Are you all right!"

"Oh, yes!" he got himself to answer, shakily.

"Do we have a comet?" Right at this moment, the opening chords of the final movement of Beethoven's first began.

"Do you three ever have a comet!" I repeated Jim's words as he explained that his telescope was showing at least several nuclei, all connected by dust, five tails, and two long pencil-thin lines (that turned out to be trains of dust) on either side of the nuclei—all as the music lurched into the movement's opening crescendo of what we renamed Beethoven's Comet Symphony. Later, Jim reported to the Central Bureau for Astronomical Telegrams:

> It is indeed a unique object, different from any cometary form I have yet witnessed. In general, it has the appearance of a string of nuclear fragments spread out along the orbit with tails extending from the entire nuclear train as well as what looks like a sheet of debris spread out in the orbit plane in both directions. The southern boundary is very sharp while the northern boundary spreads out away from the debris trails.[2]

At Palomar, the energy in that little room was extraordinary. We knew that our program had uncovered the most unusual-looking comet in history. Comets typically sport heads and tails; this one was a line consisting of several heads and tails. We were walking on the ceiling. In Gene's exhilaration that night, he had no idea that this comet's performance was only the overture, and that the comet's four-billion-year-long journey through the solar system was about to end in the most spectacular collision ever seen in the solar system. The coming event would grab Gene's lifework, confirm its significance, and blaze it over the front pages of the world.

CHAPTER 2

Of Family and Eˣ: 1925–1948

The jewel that we find, we stoop and take 't,

Because we see it; but what we do not see

We tread upon, and never think of it.

—SHAKESPEARE, Measure for Measure, *1604*

H E DIDN'T HAVE a date for his senior prom, but in the small town of Torrenson, Nebraska, that was not a problem for the tall, young high school graduate. The year was in its spring, as was this incredibly strong, gentle man named George Shoemaker. A man of many talents and interests, George never forgot his childhood view of Halley's Comet. "We had a large bay window. I remember the night; it was a really clear Nebraska night, and the tail dragged clear across the sky. That is a memory that stayed with me all my life."[1]

The senior prom was a number of years later, the day the Scotts dropped by. These old family friends were Shoemaker neighbors for as long as George could remember. In fact, young Muriel May Scott, about a year and a half older than George, claimed she had given the boy his first walking lessons when they were neighbors in the Nebraska town of Franklin. As time passed, the Scotts decided to leave for the West Coast to find a college for their daughter Muriel, and of course they planned to stop, on the way, to visit their old friends the Shoemakers.

When the car pulled in, George went outside to handle their luggage. He took one look at sixteen-year-old Muriel, rushed inside with the bags, and announced, "Mother, I'm going to marry that girl!" That evening, George took Muriel to the senior prom, and true to his word, later married her.

George's first years at college were cut short by phone calls from parents who did not appreciate his need to complete his education. The young man would rush home to assist his parents in one way or another; one college try was cut short by a near fatal motorcycle accident in which he fell on his head. Comatose for more than a week, he regained consciousness and slowly recovered. The next fall, 1927, he enrolled at the University of California at Los Angeles, where he took advantage of his strength and agility to major in physical education. Meanwhile, that same fall Muriel became pregnant and had to suspend a promising teaching career. At the time, teachers were forced to take a leave of absence for the full term of a pregnancy. On April 28, 1928, she gave birth to a son, Eugene Merle Shoemaker. Meanwhile George's eligibility to play football expired just as his son was born, and so the family headed north to Eugene and the University of Oregon. George signed up illegally for football there, cagily using his middle name Estell. The ploy worked for a few weeks, until a game in the fall of 1929 when a former teammate spotted him in a line of scrimmage. "George," he inquired, "what are you doing here?" The college football career came to a crashing end.

After George completed his undergraduate degree, the family relocated to New York City, where they, along with George's sister Alice and brother-in-law Jim, shared a walk-up flat. At Christmas 1931 the family watched out the window as a Santa Claus parade marched along the street below. "Gene," Jim called to his precocious nephew, "come see Santa Claus!"

"No," the three-year-old replied.

"Why not?"

"Because there isn't any Santa Claus."[2]

A few weeks later Muriel began teaching at the School of Practice at Buffalo State Teacher's College (now the State University of New York at Buffalo). Instead of today's usual practice of sending teachers out to public schools for their student teaching, this college had its own school. Muriel was able to get her son registered

1. Gene and his sister Maxine, 1937.

at the school's kindergarten, even though he was three months shy of the school's minimum entrance age of four.

Gene's youth was punctuated by a fairly complex series of moves and family separations. His father was desperately unhappy living in both New York or Buffalo, cities where he was unable to find work except for a briefly held position in 1937 as a physical education teacher in Jamestown, south of Buffalo. Yearning to return to the wide-open spaces of southeastern Wyoming, George and his family moved to Laramie, where George landed a job at a Civilian Conservation Corps camp as education director. He also bought a farm to the east, on Wyoming's North Platte River in the Goshen Hole region near Torrington. George loved farming, and did very well at it, nearly paying off the mortage with his first year's crop of beans. However, Muriel hated the primitive farmhouse; both she and Gene's new sister Maxine had health problems there. Sorely tempted to leave when the teaching position at Buffalo's School of Practice was offered to her again, Muriel suggested a happy separation so that she could teach, rejoining her husband each summer. She accepted the job and returned to Buffalo with Gene and Maxine. Gene entered fourth grade at the Practice School in Buffalo.

FIRST COLLECTIONS

In 1935, in the middle of this confusing period, Muriel gave her seven-year-old son her father's set of marbles. Fascinated with this beautiful collection, which included some specimens of agate, Gene began to search the neighborhood collecting interesting rocks wherever he could. The following summer, during a trip with his father to South Dakota's Black Hills, Gene was so taken with the rose quartz and other minerals in the area that he gathered samples of them; by trip's end he was accumulating them furiously.

By the time Gene entered fifth grade, the Buffalo Museum of Science had begun a program that was highly enlightened for its time, and ours. It involved evening classes in sciences as diverse as geology, mineralogy and stone cutting, botany, and aquatic biology. The course even used college-level textbooks. Muriel urged

Gene to register for this advanced program, and he thoroughly enjoyed it. His group went on field trips to a place south of Buffalo called Eighteen Mile Creek, where Gene reveled in the rich trilobite collections in the Devonian rocks.

For several summers the Shoemaker family headed west from Buffalo for Torrington, Wyoming, where they passed the cool summers with George's parents on the north shore of the North Platte River. Although it was a vacation, Muriel insisted that for her children it be a productive one. Anxious to improve her son's poor reading skill, Muriel often took him to the library. She gave him the choice of what he wanted to read there, and he concentrated on geology and archaeology. Gene also learned to read music from a local teacher who gave him violin lessons on summer afternoons. Gene particularly enjoyed the outdoors. "The streets were paved with pebbles from the North Platte River," he remembered,[3] and he collected rocks and pebbles from the neighboring banks of the river. By the end of the summer, Gene had filled a three-pound coffee can with his specimens, which included some moss agates as well as Wyoming jade.[4] When the time came for the cross-country drive back to Buffalo, he wanted to take his rocks with him and did not understand his mother's protest that the car was too full, with four people and their luggage, for even the slightest addition. Just before they left, Gene's grandmother quietly found a hidden spot for the coffee can, so Gene's collection made it back to Buffalo.[5] "Those trips just pitched it for me," Gene recalled. This perfectly normal young boy was on his way to becoming a scientist.

A BEE MOVIE

Gene was also on his way to becoming capable of some mischief. One day, when Gene was about ten, he scratched his own name in a bathroom mirror in his elementary school; one wonders why he used his own name! And one lazy summer afternoon, the family went to the movies in Torrington. Unknown to any of the others, Gene thought he'd try an experiment. Having collected several bees in a jar, he brought them to the theater. He wanted to open the jar and watch the reaction of the crowd as the bees flew out, attacking everyone at will. Sitting smugly in his chair, he imagined the consternation of the movie patrons and the theater staff as their afternoon's entertainment was interrupted by angry bees flying about the theater. Gene waited for a suitably dramatic moment in the movie, then carefully opened the jar. Nothing happened! The bees were so subdued that they just sat there! Gene sat in his chair with the jar, quietly shaking it to encourage its contents to fly out and wreak havoc. After some shaking of the jar, the bees finally flew out. One or two patrons panicked and fled, but Gene was on the whole quite disappointed with his test.

Gene also enjoyed sharing ghost stories. Making them as frightening as possible to maximum effect, he often regaled his young relatives with twilight tales of mummies. One night he announced in his deepest voice to his younger cousins, Betty and her five-year-old brother Dorsey, that a mummy would spring to life and steal Betty from her brother. Dorsey was terrified, and Gene had to spend much of the rest of the night trying to persuade the younger child that it was only a ghost story.[6]

This diabolic side to Gene's personality, as well as the swashbuckler aspect that so many knew him for in later years, obviously came from his father. During one of his several jobs, as a grip in a Hollywood studio, George rather enjoyed being a strikebreaker, coming to work armed with ear syringes filled with ammonia. He would squirt the ammonia at people coming to challenge him. On another occasion George was driving along through Nevada in his green Dodge, on his way home from college. The road was deserted except for one other driver moving slowly along the darkened road. George passed the car and continued on, but the other driver then sped up, passed George, and then slowed down. George went along with this game through several passes. Finally he cut the other driver off and forced him to a stop. George silently approached the front end of the other car, lifted it effortlessly, and set it down in the ditch by the side of the road. Then he calmly went to the rear of the car, picked it up, and set it down in the ditch as well. Rubbing his hands together, he returned to his car and drove off without a further word.

Gene's personality also had a decidedly perfectionist aspect, or "Caltech side," as his wife Carolyn later called it. He inherited this part of his nature from his mother, who was a very organized woman. "Nothing," sister Maxine remembers, "interfered with her organization; you could set your watch when she went out to hang sheets." Muriel was also superb at drawing, a skill she used to advantage both in her teaching and with her children, and one that Gene inherited.

High School

Gene stayed in the Buffalo School of Practice until 1942, when he graduated from ninth grade. Christmas 1941 was a sad time; with war having just been declared, Gene's father could not catch a train to Buffalo, where the rest of the family spent a beautiful, white Christmas.

By the spring of 1942, it was time for the family to reunite. Just before they left, fourteen-year-old Gene had an emergency appendectomy; it might have been prudent to change plans to leave for the West Coast but with travel so limited and difficult during wartime, Muriel dared not cancel her reservations.

Finally, the family was together again, now in Los Angeles. Gene brought his growing love of minerals to Fairfax Senior High School, where he embarked on an accelerated three-year curriculum. While there he joined the school's gymnastics team, which helped give him his later goatlike agility as a field geologist. Fairfax was in the heart of a rapidly growing area of Jewish families, most of whom had settled there from New York. The school was more than 90 percent Jewish, a fact not lost on Gene when he showed up at school one Yom Kippur and was shocked to find the building all but deserted.

As a school for some of California's brightest students, Fairfax gave Gene the opportunity to become active in several activities unrelated to science. Still enjoying the violin from his summers in Wyoming, Gene played respectably for his school orchestra's performance of *The Merry Widow* and found that experience an effective way to develop a lifelong love of music. Based on the 1934 film starring Jeanette MacDonald and Maurice Chevalier, the story follows a lively widow who owns virtually all of the small European country of Marshovia. When she moves to Paris, Marshovia is threatened with financial ruin, until Captain Daniel arrives in the city of lights to bring her back and save her homeland. It would have been priceless to see Gene and his violin playing *Merry Widow Waltz*; *Girls, Girls, Girls*; and especially, in view of his later career as an astronomer, *Tonight Will Teach Me to Forget*.

During his high school years Gene also enjoyed his first job. It was part-time work in a plastics factory, from which he presented his mother with one of his first products, a pin with the letter *M* emblazoned on it. Gene's increasing interest in rocks led to a summer job working as an apprentice lapidary at the end of his first year in high school. He enjoyed cutting and polishing so much that his parents worried he might skip college to become a professional lapidary. Gene's happiness with his lapidary work, coupled with the fact that he did not work terribly hard at Fairfax, worried his parents. How would their son ever turn out, they fretted, if he never went to college?

E^x!

And seeing ignorance is the curse of God,

Knowledge the wing wherewith we fly to heaven.

—SHAKESPEARE, Henry VI, *circa 1590*

In the fall of 1944 at the age of only sixteen, Gene Shoemaker entered Caltech. It was a critical time during World War II, and Caltech's normal annual routine was shifted to three semesters each year, beginning in the fall, midwinter, and summer, with no period for summer vacation. Navy engineers needed training as quickly as possible, and civilian students on campus followed the same schedule, most of them heading into the service after their first year. Gene wanted to do that as well, perhaps as a specialist in electronics, but the military regulation was that he could enlist only with permission from his parents. George wasn't in favor of his son joining the military at that time. The older Shoemaker believed, correctly, it turned out, that the war would be over within a year, and that the best thing for his son was to remain at Caltech. "I respected Dad's knowledge and his view of things," Gene said. "In fact I was overawed and overwhelmed by his knowledge of the world. It would have never occurred to me to argue with him."[7] Gene remained a student, despite the geology department being virtually defunct, its senior staff away with war-related work. That wasn't too much of a problem for Gene; he had completed his introductory courses in physical and historical geology, and he now concentrated on completing his other course requirements.

Gene also joined the Caltech cheerleaders. That enterprise, important for a school that had mostly men students at the time, was a tremendous release for a young man with a lot of energy to spare. Caltech sponsored a football team with the minimum number of players; without any relief, the team tended to do well in the first half of a game but would lose energy at the end. The team's home games were played in the Rose Bowl stadium, with cheerleading from Gene and another student. Gene did his act by enthusiasm alone, with neither special costume nor special training. He developed his own routine and practiced it in his room on campus. Years later he gave me an example of his prowess, a takeoff on elementary calculus where e is a fixed value in calculus, and x is an unknown:

> *e to the x the x the x*
> *e to the x the x the x*
> *e to the x the x the x*
> *e to the x dx [delta x]*
> *Sliiii . . . de rule! Tech tech tech tech tech tech tech!!!*

The key to the cheer was that the opposing team had no idea what it was! Nor did they follow:

Cotan, tangent, cosine, sine!
3.14159!
Sliiii . . . de rule! Tech tech tech tech tech tech tech!!![8]

By the fall of 1946 and the beginning of Gene's senior year, the geology faculty had reassembled and Gene made up for lost time; one of his professors later remembered him then as "a feisty, very smart, young feller."[9] He needed to be: his final three-semester year was pure geology, with half again more than a full load of courses so that he could graduate. Visible in virtually every geology course that year, around the department he was called Supergene, a pun on a geologic term applied to ore minerals formed when surficial processes lead to ore deposits. (If processes at depth lead to ore deposits, by the way, they are called hypogenes). "Supergene was arguing for a hypogene origin of the Colorado plateau," recalls Lee Silver about Gene's later work there. Since it usually turned out that Gene was right, his colleagues nicknamed him Supergene. "I squirted out of Caltech in two and two-thirds years; the last year the navy program was over so I was in my senior year—I was barely nineteen when I graduated." Gene took virtually all his geology courses in his final undergraduate year. Being so far ahead of his age, the geology department encouraged him to "consolidate his gains" and take a year's graduate work at Caltech before moving on to another university for his doctoral studies. In the spring of 1947 he took a tough civil service examination. Barely passing it, he was offered a job but declined it in favor of remaining at Caltech to work on his master's degree.

That fall, Gene began his one-year master's program as well as a long-term acquaintance with Professor Robert Sharp. Although Sharp found him very energetic and enthusiastic, Gene seemed somewhat naive and uncertain in his interactions with other students. Sensing that Gene was "a kid who liked exploration and challenge," Sharp suggested a mapping project in an area west of Los Angeles along the Santa Clara River. That fall Gene spent more than a week in the field starting the project, but returned unimpressed with its possibilities and decided not to complete it. This early incident echoed a later tendency for Gene to focus enthusiasti-

2. Gene reaches eighteen.

3. Gene, a senior at Caltech, 1947.

cally for a while on an effort, only to drop it if he did not see it heading anywhere.

The following spring, "the cobbler"—his classmates' play on his name—finished his master's. However, his education still lacked sufficient fieldwork. Caltech's normal undergraduate requirement was a year's introductory field course, a year's advanced course, and two summer field camps lasting six weeks each. Although he had completed the first course during his final undergraduate year, a summer camp after graduation, and the advanced course during his master's year, Gene still lacked the experience of the second camp. The solution was that he spent the summer of 1948 as a teaching assistant for Caltech's field camp. The fieldwork took place high in the Zuni mountains of northwestern New Mexico. "What a relief from the California coast," he recalled. "Here we walked through a Ponderosa Pine Forest—I thought I had died and gone to heaven!"

Gene was enjoying this Caltech camp not far from the small New Mexico town of Gallup. Although he didn't know it at the time, someone born in that town—Carolyn—would play a key role in his later life. In any event, the camp marked the end of the first phase of Gene's association with Caltech. When the camp ended, Gene and his friend Bill Muehlberger joined Dick Jahns, the camp's faculty mentor, on a mapping project, also in northern New Mexico. After Jahns laid out the grid, the two younger geologists developed a topographic base map and began plotting onto the map the interesting Precambrian metamorphic geology. The pair used an aneroid barometer for the purpose of measuring the elevations of the various rock outcroppings above sea level. The instrument measured subtle differences in air pressure relative to a reference base. By the end of the first day the geologists dropped the first barometer, and the spare one was destroyed in a fall the second day. For the rest of the fieldwork the team had to rely on Brunton compass angles and trig tables.[10] Always a geologist's companion, the Brunton compass was designed for field mapping in geology.

Early in 1948 Gene accepted a position with the U.S. Geological Survey. His job involved geological mapping in a search for deposits of uranium in western Colorado. At the dawn of the nuclear age, the United States had very limited knowledge of where the uranium deposits were. Headed by geologist Richard Fisher, this project was vital both for the production of nuclear weapons and for the fledgling nuclear power industry. Those were the days when we were told that nuclear power would be too cheap to measure—but the nation needed a supply of uranium first.

CHAPTER 3

Over the Sea, Over the Sea: 1948

Lady Moon, Lady Moon, where are you roving?

Over the sea, over the sea.

Lady Moon, Lady Moon, whom are you loving?

All that love me, all that love me.

—Nursery Rhyme

LATE IN THE EVENING of April 28, 1948, Gene's twentieth birth-day, the Moon rose in the southeastern sky. A few days past full, it shone beautifully on the Colorado plateau near West Vancoram, where Gene was assisting Richard Fisher prepare for the diamond drilling project that would search for a badly needed supply of uranium ore. Could Gene's single look at that rising Moon have planted a seed in his mind? In any event, that seed germinated one morning that spring as Gene drove the five miles from West Vacoram to Naturita, the headquarters of the Vanadium Corporation of America, for his breakfast. Suddenly the thought hit him: "I want to go there! I want to be one of the first people on the Moon. Why will we go to the Moon? To explore it, of course! And who is the best person to do that? A geologist, of course! I took the first fork that went to the Moon that morning."

There would be other forks in a road to the Moon that would last the rest of Gene's life. There would be unpleasant lessons learned about why a nation would really want to head for the Moon. But on that morning, the Moon beckoned. It was an uncertain place, its surface carved from forces scientists were just beginning to understand. A look through the smallest of telescopes reveals that craters are the dominant feature on the Moon. But how did they get there? The following year, Ralph Baldwin would publish a logical argument that the Moon's larger craters were formed by impact, but minds were slow to turn and many students of the Moon continued to hold that the craters were mostly volcanic in origin.[1] Gene set about to study those aspects of geology, like volcanoes that erupt in violent explosions, that would help determine whether the craters were volcanic or the result of collisions.

For Gene the thought of walking on another world, and of doing geology on a different world, was thrilling beyond description. It was also so unorthodox that he dared not share it with anyone. At the time, the United States was testing single-stage rockets at White Sands Proving Grounds in New Mexico. There was reason to be optimistic: from a Caltech newspaper, Gene knew of White Sands, home of a cache of V2 rockets rescued from their wartime German base at Peenemünde on the shore of the Baltic Sea. Werner von Braun, master of America's fledgling space program, and other scientists were now launching V2s from White Sands. It would be a year before the first two-stage rocket, the "Bumper," would fly successfully to a height of 260 miles above the New Mexico desert, after three failed attempts. Clyde Tombaugh, the astronomer who years earlier had discovered the solar system's ninth planet, was in charge of the optical tracking of the mission at White Sands. "The path was so high," Tombaugh said, "that the jet spun out like a comet's tail. It was the most magnificent thing you ever saw." Although this two-stage rocket got close to launching a satellite into orbit, a third stage was needed to turn a payload into a new artificial moon. And an orbiting satellite would be only the first small step toward bridging the 238,000-mile gap between the Earth and the Moon. Meanwhile von Braun and Tombaugh tried to make the dream a reality by considering a third stage that would send a satellite into orbit. Called Project Orbiter, the plan seemed a shoo-in as a second step into space after the success of the V2 program.[2] But the Pentagon directed that there would be no third-stage rocket, no Project Orbiter, and no artifical moon program. "There are harder fights fought in the Pentagon," Tombaugh noted wryly, "than are fought on the battlefield!"[3]

Thus, in the spring of 1948, going to the Moon seemed the dream of science fiction writers and a single young scientist named

Gene Shoemaker. The dream would have to remain a dream, a foundation whose stones he would lay, one at a time, as he built his geological career. This was Gene's plan, but his job under Survey scientist Richard Fisher in the search for uranium ore—was down a rather different path.

For a small group of scientists living in diverse locations, the seed of space exploration had already germinated. Time and the Cold War would later cause it to flower.

CHAPTER 4

Springtime, Carolyn, and the Colorado Plateau: 1948–1952

Love, whose month is ever May,

Spied a blossom passing fair. . .

—SHAKESPEARE, Love's Labour's Lost, *circa 1597*

IN THE SPRING of 1948, a young chemical engineering major named Richard Spellmann graduated from Caltech. In the audience was his proud mother, and after the ceremony ended, she met her son's roommate. When she returned to her home in Chico, a town in the California orchard country north of Sacramento, she excitedly told her daughter, Carolyn, about what she had seen and done. Included in her stories was frequent mention of Richard's roommate, a nice young man named Gene. Suspecting that her mother was trying to make a match for her, Carolyn listened politely, but paid as little attention to her mother's stories as she could get away with.

Two springs later, Carolyn heard about Gene again. This time, Richard was getting married, and Gene was about to arrive in Chico to be his best man. The pressure on Carolyn was increasing. Not only was Gene making a special trip out there, but here also was a chance for Carolyn to meet him in person! Dating someone else at the time, Carolyn still resisted, until, in a valiant effort to get his sister to look at Gene with an open mind, Richard bribed her. If she would only spend some time with him, his fianceé, and his best man in the days before the wedding, Richard would give his sister a table cloth from Guatemala. Sounding like a good offer, Carolyn accepted, although she admits she would have gone along with the plan even without the gift. But not before she teased her brother a bit: The day before Gene was to arrive, Carolyn answered the phone. When Richard asked who the caller was, Carolyn replied that it was Gene; that his grandmother was ill and he would not be able to come to the wedding. "I didn't even know he had a grandmother," Carolyn noted. "I did not have any living grandparents." Her brother was disappointed, and somewhat puzzled, by this report.

Meanwhile, the real Gene was heading west toward Chico from his home in Grand Junction. As he drove into Nevada in "Jezebel," his old Ford, he stopped for a round or two at the roulette wheel, and by dinnertime that evening he arrived at the Spellmann house in Chico. Carolyn met him at the door and remembers him wearing Levi's, an engineer's cap, a mustache, and a big grin. He's a little like Richard, she thought.

During the next few days Richard; his fiancée, Doreen; and Gene spent some time integrating Carolyn into their foursome, spending time at places like Mount Lassen, a favorite family playground. After the wedding, Richard and Doreen left for their honeymoon; Gene decided to stay for a few days longer.

It was a curious but invigorating time. When Gene invited Carolyn for a drink the evening after the wedding, Carolyn thought he meant sodas. Gene was a little surprised but laughed heartily when they arrived at a drive-in place that served soft drinks! They talked about lots of things that evening, and after they returned to Carolyn's house, Gene passed the evening showing Carolyn a trick or two. In one, he stretched out on the floor, placed a nickel on his nose, and wiggled his nose until the coin fell off. "He'd just lie there and wiggle his nose," Carolyn recalls. "He could do it very quickly; I tried it but couldn't do it." Next, Gene displayed his musical prowess by running his finger around a crystal glass, producing a tone. He tried other glasses and produced different tones, each one unique to a particular-size glass.

Next morning, Gene climbed into Jezebel for his return drive to Grand Junction, and to his uranium survey with Richard Fisher. The August morning was very hot and grew hotter as he drove east. Jezebel did not do well in that weather and, on the way, broke down from vapor lock. Gene pulled over, got out of the car, and,

4. Gene with his trusty Ford, Jezebel.

5. Carolyn's brother Richard, Gene, and Carolyn's father.

in a display of the quick temper he would later become known for, slammed the door so hard he broke the window. Gene wrote up this episode in a long letter to Carolyn, who thoroughly enjoyed the details of Gene's narrative. At the end of summer, Gene headed east for graduate studies at Princeton.

PRINCETON

Princeton, New Jersey, in the fall of 1950 was a serene place. The home of some of the most brilliant minds of the twentieth century, including Albert Einstein, it offered a studious and challenging atmosphere to graduate students but lacked, as Gene was wont to complain in letters to Carolyn, the field-and-frontier Western spirit of Caltech, and certainly the informality of the field near Grand

Junction. It took a little adjusting for Gene to get used to his new environment; unlike the overachieving Caltech students, the ones working under Gene at Princeton seemed more cerebral. As Gene set out to study at Princeton, his first interest centered on how certain rocks form and alter under heat and pressure—a study of igneous and metamorphic petrology. He looked forward to working under Arthur Buddington, one of the nation's top petrologists. However, his early weeks at Princeton were undemanding, with course work and research, so Gene took advantage of the time to pass his language requirements in French and German. At the time, the language requirement was a specific one, to provide the ability to keep up with the geological literature in other languages. Accordingly, the geology department saw that this requirement was met through its own resources and not via the university's language departments. When Gene arrived at Princeton, he bought a copy of Eduard Seuss's *Le Face de Terre*, a famous turn-of-the-century introduction to the Earth with editions in German as well as French. After reading a few pages of the French version, he sought out the department's French expert, an invertebrate paleontologist named Benjamin Howell, to pose a routine question about the details of the French language requirement. Howell rose from his desk, fetched a book written in French, and handed it to Gene. "Translate this!" Howell ordered. Gene began reading the book's opening pages, translating the sentences into halting English. Howell provided a difficult word here and there. In about twenty minutes Howell asked Gene to stop, smiled, and said that the young student had just passed his French requirement! Not bad! Gene thought. I wonder if German will go as fast. Gene then began reading the German version of the same book, *Das Antlitz der Erde*. Although he eventually passed German, it took him longer than twenty minutes.

Meanwhile, Gene kept Carolyn up to date with letters about his progress at Princeton. Although Carolyn enjoyed answering Gene's letters at first, she soon concluded that since she probably wouldn't ever see him again—at least not for a long time, while he finished his dissertation—she shouldn't invest any more time in continuing the relationship. So without further thought, Carolyn stopped answering his letters. Although this seemed practical to her, Gene immediately noticed the absence of letters and wrote to ask why she had stopped corresponding. Surprised, and impressed that he cared about her more than she had thought, she answered that particular letter and started writing to Gene more regularly. Gene wanted to return to Chico in the summer of 1951. He now wrote with an invitation: would Carolyn and her parents wish to join him on a camping trip through some magnificent Western scenery. Carolyn's father had suffered a heart attack earlier that year and decided not to accompany the trio on this trip.

Gene returned to Grand Junction from the weekend in Chico, and continued his work there until Carolyn and her mother arrived for their vacation. At the rendezvous they quickly decided to use the larger Spellmann vehicle rather than risk the temperamental Jezebel for the trip. The three headed on a camping trip through Montana to enjoy the spectacular scenery of the Colorado Plateau. During the trip Gene was entirely in his element, interpreting the geology of these magnificent vistas. For Carolyn, it was an eye-opening experience. She had taken a course in geology at her college in Chico after World War II, but that experience had turned her away from the science. The professor was good at writing notes on the blackboard with both hands simultaneously, telling some reasonable jokes, but then repeating the same jokes during the course's second half. Now, in the middle of Earth's majestic classroom, Gene explained the Earth in a way that filled her with wonder. "Gene told us how Earth was formed"—Carolyn described her new teacher and friend—"and what its different characteristics were." After the first week, as they headed back to Grand Junction, Gene impressed Carolyn and her mother with his agility at routinely repairing a broken fan belt and dealing with various other car problems. Once they got back to the town, Gene put Carolyn and her mother in a hotel, while he stayed in his boarding house. These few days were a sort of waiting and guessing game. "Each night Mother would say, 'did he ask you?' On the second night in town, Gene asked, "How would you feel about being married to me?" Carolyn was not sure if that was a proposal or not, but she replied, "It would be very nice being married to you!"

When Carolyn's mother asked her again, Carolyn replied that from the way Gene had proposed, she wasn't sure if she had been proposed to. As the trip went on, neither Carolyn nor her mother said anything about it. "I think I was too shy to bring up the subject," Carolyn says, "and so was Gene." As it turned out, that *was* Gene's proposal. Finally Carolyn's mother surreptitiously mentioned something about wedding plans, and Gene readily joined in the conversation. It was only at that point that Carolyn and her mother knew that Gene and Carolyn were actually engaged to be married!

Later that summer, Carolyn was visiting Gene in Grand Junction when her fiancé said, "Carolyn, there's someone here I want you to meet!" Gene's father, George, had caught up with him in Grand Junction. George tried to keep an eye out for his son and would often show up at unexpected times and places; Carolyn remembers that he wanted to look out for Gene without being overprotective.

On that evening, Carolyn didn't know who wanted to meet her. She walked to the lobby and standing there was Gene, and the unfamiliar figure of his father, who popped his first question: "Aren't you two married yet?"

August Wedding

The course of events moved quickly after that. The couple selected August 18, Richard and Doreen's first anniversary, as their wedding date. But Chico summers are very hot, and to get away with a wedding at that time of year, they scheduled it for the morning. Gene drove Jezebel out from Grand Junction, arriving in time to take a shower and ask Carolyn to iron his shirt: ironing was a new experience for the bride.

August 18, 1951, dawned clear and incredibly hot. By the time the nuptials had begun, the temperature had already soared to 118 degrees! They were married in Reverend Gould's Methodist church, and although Carolyn rather expected her groom to wear his engineer's boots, he did bring black shoes for the occasion. At

6. Wedding Day!

the time, Gene and Carolyn did not share the same religious ideas; Carolyn had been active in the church's youth groups. Gene compliantly memorized his vows, but when the question had come up during the planning about including a statement of religious commitment, Gene had said, "I can't do that, Carolyn."

The ceremony was elegant and moving. After the wedding there was a simple reception in the church's recreation room. During the reception, Carolyn's mother and Gene's best man, Jim Duderstadt, excitedly pasted Jezebel with traditional "Just married" paraphernalia. By noon, Gene and Carolyn were set to head out for their honeymoon on the shores of pristine Lake Tahoe, high in the Sierras. Imagine the best man's chagrin when the groom ushered his bride to his unmarked *parents'* car and rushed off to Tahoe! Meanwhile, the groom's parents hopped into Jezebel, waved, and drove happily off to Los Angeles.

The happy young couple spent their first honeymoon night at a ski lodge and then headed south. Gene began the trip with a promise, Carolyn recalls with a smile, that he would not mention anything geological during the entire honeymoon. This particular marital vow lasted until the couple left Tahoe and began their descent through the spectacular uplift geology of Mono Lake on the eastern Sierra slope in Nevada. This is one of the world's most dramatic examples of exposed geology, and Gene excitedly became a geologist again, explaining it to his bride.

From summits to deserts, the honeymoon trip offered Carolyn views she had never seen before. Entering the busy Los Angeles basin was a different story, however, for a couple who would always prize their wide-open spaces. They spent a few days with Gene's parents, and having traded cars once again, returned to Grand Junction.

A Marriage Begins

The Shoemakers began their life together in another young couple's home; the owners were away, so Gene and Carolyn had this first house to themselves. As in most marriages, the first year set the stage for the couple's life together. Gene would often bring home fellow geologists, former students, and other friends; when he was single these friends had entertained him and he enjoyed the chance to reciprocate those early favors. As a single young man, Gene had spent more of his time in the field than he did now, but the enthusiastic young geologist was anxious to bring his new wife into the field aspect of his career. Considering how involved Carolyn would later become in her husband's fieldwork in both land and sky, this was a wise decision. In the fall of 1951, the couple visited the La Sal mountains near Moab, Utah, and the spectacular Sinbad Valley across the Colorado border. The complex and subtle geologic unconformities in that valley—erosional surfaces that separate younger from older layers of rock—made it one of Gene's favorites, and a subject of his thesis.

Trouble on the Plateau

By this time Gene had completed the brunt of his dissertation fieldwork on the Fisher Valley–Sinbad Valley salt structure on the Colorado-Utah border. He fully expected to return to Princeton with his new bride in the fall of 1951, write up his dissertation, and finish his Ph.D. As Gene prepared for Princeton, however, he and his young colleagues were asked to attend a meeting with senior staff members of the USGS. Uranium—fuel for nuclear weapons and power—was the hottest topic at the Survey of the day, and the USGS was committed to learning as much as it could about where the uranium was, and how it got there. The project's field leader, Richard Fisher, had his own ideas on the origin of the uranium, and USGS senior brass suspected that there was some bias in the field reports. A somewhat clandestine meeting took place in Fisher's absence; Gene came to that confab with his own view, and as a result, he was given carte blanche to run his own program. Not wanting to pass up this opportunity, Gene delayed his return to Princeton. Instead of moving east, the couple bought their first house, in Grand Junction. "Gene was a major voice in the debate on the genesis of the Colorado plateau Uranium Deposits," says

Lee Silver, who knew Gene from those years. He did not follow the main line. "Gene raised questions about the conventional idea—not vicious or mean."

Several months later, Gene's parents joined them for their first Christmas. It was a family event, though Gene still had to learn a little about family. On Christmas day, Gene thought he could get away from the festivities for a while by saying he had to finish a map he was working on, but when Carolyn objected, he obeyed and rejoined the family.

New Year's Day 1952, was the beginning of one of the young couple's happiest years. There was a lot of fieldwork, especially that summer. On a typical field day, the couple would rise with first light, even though Gene was not otherwise a "morning person." As he picked up camp, Carolyn would prepare breakfast. Never talkative early in the morning, conversations would be light until fieldwork began in earnest. Around noon they would break for a simple lunch of sandwiches made on the spot. In summer, afternoon fieldwork would continue until six or seven o'clock, when shadows became too long to determine colors accurately. "It was a great way to get to know each other," Carolyn remembers. "It was like a continual honeymoon, just the two of us. We seldom pitched a tent; instead we'd sleep out under the stars." It was a habit they would continue throughout their lives. With the spectacular scenery of the Colorado Plateau as a backdrop, this honeymoon would be the envy of many a couple.

Fieldwork was a way to learn as much about each other as about the rock structures and geological history they had come to study; in fact, Carolyn often said that the best way to get to know another person is to share camping or observing. "Gene was half person, half jeep," says Carolyn. "He knew so well how to drive the primitive roads and to repair the jeeps we used." The road across Utah's Monument Valley was quite primitive and difficult to navigate. With lower speeds, there was more time to take in the spectacular scenery. It was a special and satisfying time.

That summer the couple also worked south of the Nevada town of Battle Mountain under the leadership of Jim Gilluly, a central figure in geology at the time. Although then in his sixties, Gilluly had such an incredible reserve of energy that the younger geologists were hard put to keep up with him. Not used to working with a leader as energetic as Gilluly, Gene saw the task of keeping up with him as a personal challenge. Gilluly was thorough. He taught his assistants not to sample a few exposures but to find and study every possible outcrop. In this thorough way, their reward would be a more complete picture of a particular formation and the entire structure. Gene tried to keep up not only with Gilluly but also with one of the other assistants who had similar competitive ideas. He ended each day exhausted, but he'd proved he could keep up.

Gene was also impressed with Gilluly's emphasis on doing careful fieldwork and striving for perfection, values which matched well with the teachings Gene had from his mother. Summer evenings were spent talking around a campfire about what older geologists had done and what the younger ones hoped to accomplish. When the summer's work was cut short by a late-summer snowstorm, Carolyn, now pregnant with their first child, and Gene returned to Grand Junction.

FIRST VIEW OF COON BUTTE AND ITS CRATER

The summer closed with a small side trip that, unknown to them at the time, turned Gene's career spectacularly. Gene and Carolyn were completing fieldwork in the Navajo reservation of northern Arizona with colleague Bill Newman. They left Kendrick Peak and soon found themselves not far from Winslow, and Gene suggested that they try to head for the large structure called Meteor Crater atop Coon Butte. The crater is named for the post office the federal government established near the site in the 1930s. The name Meteor had nothing to do with the crater but for the meteorites in the surrounding area. At this time Gene had little idea that their next stop would focus the direction of his career, and of our understanding of how our solar system works.

The origin of this 1.2 kilometer–wide hole in the ground was very much a mystery. The crater was owned by the family of Daniel Moreau Barringer, a mining engineer who was convinced that the crater was of cosmic origin. But most geologists concurred with Grove Karl Gilbert, a highly regarded turn-of-the-century geologist who had concluded that the crater was the result of a volcanic steam explosion, a logical suggestion since the crater is near a known volcanic field with diatremes, or volcanic vents drilled explosively through the enclosed rock. Gilbert had hoped that a huge mass of iron meteorite beneath the crater floor would cause an anomaly in his compass reading. But Gilbert's magnetic test showed no significant deflection, and so he suspected instead that it had to be volcanic. Gilbert made a philosophical issue of his study, featuring his work in his retiring presidential address at the Geological Society of Washington, but from a philosophical point of view. The "Origin of Hypothesis" paper discussed how a hypothesis is born, figured through, and then either accepted or discarded, using the crater in northern Arizona as an example. "The mental process by which hypotheses are suggested," he wrote, "is obscure. Ordinarily they flash into consciousness without premonition." Gilbert proposed the idea that hypotheses come out of analogy: to explain an unusual event or a peculiar feature, the scientist looks to see what aspects of it might have been explained before, and builds the hypothesis on what is familiar.[1]

Gilbert compared the two favored theories of the crater, that it was formed by the explosive eruption of a volcano, or by the impact of an object from space. For craters here on Earth as well as on the Moon and other planets and moons, scientists were forever debating which giant pockmark was caused by one and which by the other; in fact Gilbert gave the first clear exposition supporting the theory that lunar craters could have been formed by impacts of other objects in space. "What would result," Gilbert asked, "if another small star should now be added to the earth, and one of the consequences which had occurred to me was the formation of a crater?" Gilbert thought that the main body would be composed of iron, and that if it lay beneath the crater it should cause "a local deflection of the magnetic needle." Since there was no deflection, he concluded that the crater was the result of the other, volcanic cause.[2]

As a geological icon studied by a geologist he admired, Meteor Crater was high on Gene Shoemaker's list of places to visit. But the hour was getting late, the group was down to pennies, and they had to return that evening to Grand Junction. They drove quickly toward the site, but since they could not afford the admission fee, they drove up an access road on the west side of the crater. They left their jeep and crawled up the hill to the top of the Butte. The Sun was already setting over the structure, and the long shadows offered a magnificent view. "The crater was an overwhelming sight," Carolyn says, remembering how they looked quietly across its 1.3-kilometer girth. They stayed just a few minutes, and then began their drive home to Grand Junction.

CHAPTER 5

A Revolution in Earth

Ring out, wild bells, to the wild sky,

   The flying cloud, the frosty light:

   The year is dying in the night;

Ring out, wild bells, and let him die.

Ring out the old, ring in the new,

   Ring, happy bells, across the snow:

   The year is going, let him go;

Ring out the false, ring in the true.

—TENNYSON, *1850*

THE EARTH is a book, its story carved for all time in pages of rock. The pages of this incredibly long story, though, are cracked, baked, and broken, and some are missing. The story can be read and understood by any person who is trained, like Gene Shoemaker, as a geologist. When Tennyson's wild bells rang out in *In Memoriam*, they were celebrating the dawn of a new understanding of the Earth and its life forms; an understanding that the evolution of Earth and its life was slow and gradual. In this chapter and the next, we deviate from our story to offer a geologic context for the work that Gene was about to perform.

In this chapter we will explore two different ways of viewing the Earth's evolution. The earlier school of thought, now called catastrophism, sees the Earth's geologic history as a series of violent events caused by forces we do not see today. The later school, which became widely accepted in the nineteenth century, is uniformitarianism—the concept that the present is the key to the past and that processes we understand today shaped the world throughout time. Today, thanks in part to Gene Shoemaker's ideas about the role of impacts through geologic time, many geologists now see the planet's history as a combination of both schools.

However much geologists have disagreed over the centuries, their work has always had the joy of being close to the land. "A child gathering pebbles on the seashore"; so Sir Isaac Newton thought of his life as a scientist. For scientists like Gene Shoemaker, Nature is an incredible coastline, inviting us to seize on whatever pebbles she throws at our feet. Ancient philosophers like Thales and Democritus forsook their inherited wealth so that they could pursue their studies.[1] Gene also had the advantage of an inquiring mind like Mozart's, one that had been directed toward its prime subject since its youth. For Mozart, that direction was music, for Shoemaker, it was the Earth.

A LOOK AT EARTH'S PAST

Knowledge about the Earth has come in small steps from the work of a long line of geologists. The ancient Greeks understood that the Earth was round as they watched departing ships appear to sink as they approached the horizon. They also understood that an eclipse of the Moon occurs when the Earth casts its shadow on the Moon. That shadow is round, so the Earth must be shaped like a ball or sphere. Some two hundred years before Christ, Eratosthenes measured the actual size of the Earth. He calculated the angular difference in the Sun's position at high noon on the day of the summer solstice in two different cities, and thus extrapolated its circumference.

How old is the Earth? Using biblical genealogies and the "begats," the ancient Hebrews believed that our world was formed some five thousand years ago. In 1650 James Ussher managed to narrow that down, somehow calculating that the Universe was created on what would have been October 22, 4004 B.C. As geologists began to peer more deeply into the rocks, they realized that the

Earth must be far older than that. Struck by the similarity of the motion and direction of our solar system, Pierre Simon Marquis de Laplace proposed, in 1796, a theory that the Sun and planets condensed out of a rotating cloud of dust and gases. As it spun, it threw off a series of concentric rings, each of which condensed into a planet. Although it resembles an idea that Immanuel Kant set forth in 1755, the Laplace theory was probably developed independently and with far greater detail. It was possible, some thought, that the nebulae that were being discovered across the sky were far-off examples of solar systems in formation.

In the nineteenth century some geologists thought that the Earth had begun very hot, and their simple calculation of how long it took the Earth to cool to its present temperature led them to think of the Earth as being eighty million years old—a revolutionary idea! In 1905, using newly discovered radioactive decay, John William Strutt determined the age of a very old rock at two billion years. In the first half of the twentieth century, R. T. Chamberlain and F. R. Moulton suggested that another star passed close to the Sun and caused two great eruptions of gas to burst forth, one on either side of the Sun.[2] The gassier planets, like Jupiter, Saturn, Uranus, and Neptune, were formed from one of the eruptions; Earth, Venus, Mars, and Mercury were formed out of the other. The problem with this theory is that it doesn't explain how the distant planets got their great angular momenta. Later, astronomers Russell and Lyttleton suggested that the Sun had been part of a double star system. While that is certainly not unusual—about half the stars we know of are double systems—in the Sun's case the double was responsible for the creation of the Earth. In this model, Chamberlain and Moulton's intruding star interacted tidally with the Sun's companion, a disturbance that resulted in the formation of the planets.

In 1952 Harold Urey, later to win a Nobel prize in chemistry, returned to a Kant-Laplacian proposal that the planets were formed slowly by the accretion of small bodies from gas and dust. Just as most scientists were now seeing the history of the Earth in a uniformitarian way, so too was this the thinking about our planet's origin. Sir Fred Hoyle, the British astronomer, postulated that the proto–solar system was a rapidly rotating and shrinking cloud of gas. When the cloud shrank to the size of the orbit of Mercury, about thirty-six million miles in radius, some gas was expelled from the disk, and expanded quickly to form an outer disk. The magnetic lines of force coming out of the central body tended to brake its own rotation, but at the same time speeded up the rotation of the outer disk. This theory would explain the rapid angular momenta of the outer planets. Near the Sun, where the temperature was high, the heavier atoms condensed giving nearby planets their rocky land. In the outer part of the system, where it was cold, lighter atoms condensed, and gaseous planets formed.

By the 1960s most geologists saw the history of the planet in this uniformitarian way, from the gradual consolidation of the planet from smaller objects, to the slow and steady buildup of continents and their features, and the evolution of the atmosphere and the oceans. On the horizon, however, were a few seemingly minor difficulties. Was it possible, for example, that historically the geologic process could have existed in greater degrees and at more accelerated rates than we now see? And could the most violent episodes of orogeny—earthquakes and volcanism—have preoccupied the early years of Earth to such a degree that changes were truly catastrophic?

Uniformitarianism versus Catastrophism

How we interpret the development of the Earth should be subject only to the scientific method, but over the centuries the interpretation was also influenced, as we shall see, by cultural bias. Students beginning their college course in geology, at least until recently, were taught to see the development of their science as a sort of a battle between good and evil, the schools of uniformitarianism and catastrophism. Leonardo da Vinci, perhaps the earliest uniformitarian, thought that the shape of the land could be interpreted in terms of everyday processes. As further developed, the concept of uniformitarianism holds that "the present is the key to the past," that the same processes that produce the natural changes we observe on the Earth today have been the only ones acting in the past, and that all large scale changes have been the result of gradual processes.

Catastrophism says that the Earth has experienced great catastrophes from time to time, and that the processes induced by these episodes have wrought great changes in short time spans. Catastrophism in its traditional sense dates from biblical events like Noah's flood, which are no longer repeated in time or in strength. The past was controlled by forces not seen today. Catastrophism built on the biblical theory by suggesting that the face of the Earth today was shaped by cataclysmic events unlike anything we now see.

Scientifically, catastrophist geologists believed that the violence of the great extinctions of the past, the upward thrusting of mountains and cutting of canyons, were events that could be explained by causes that we do not see today on Earth. Uniformitarians replied that the while these effects could be described this way, it is far more likely that they occurred during quieter episodes over a longer span of time, as we observe them happening today. Reverend Thomas Burnet, to cite an early example, published a thesis in 1681 called *The Sacred Theory of the Earth*.[3] This work made the then-new point that the biblical Noah's flood, might have resulted from natural instead of exceptional causes.[4]

By the end of the eighteenth century, geology was leaving the study and heading into the field, having rejected speculation in favor of observation and measuring. In 1778, encouraged by discoveries made in the mining region of northern Europe, the science professor Severinus prescribed an activity to his students that Gene Shoemaker would echo two centuries later:

> Go, my Sons, buy stout shoes, climb the mountains, search the valleys, the deserts, the seashores, and the deep recesses of the earth. Look for the various kinds of minerals, note their characters and mark their origin. Lastly, buy coal, build furnaces, observe and experiment without ceasing, for in this way and in no other way will you arrive at a knowledge of the nature and properties of things.[5]

Abraham Werner, a Prussian geologist, was one of the first scientists to dare to publish, at the beginning of the nineteenth century, a complete *Theory of the Origin of the Earth*. In it he suggested that the primordial Earth was covered with an ocean dense with the suspension of materials that now form the sedimentary rocks of the Earth's crust, but he gave no explanation of how this universal suspension came about. Although his ideas are now discredited, Werner dealt with the subject in such a mature and scientific manner that he is sometimes credited with being the first to give geology its status as a science.[6]

In contrast to this "Neptunian" theory, in 1788 the Scottish scientist James Hutton made a case for a "Plutonic" origin of Earth in which both fire and water played roles in forming the rock features of the crust. Hutton's greatest work, *Theory of the Earth with Proofs and Illustrations*, appeared in 1795.[7] Unlike Werner, Hutton paid attention to the fact that sedimentary beds often rest discontinuously at odd angles atop earlier beds. These unconformities represented revolutionary episodes of uplift, warping, and periods of erosion. The geological "column" (the grand structure of rock upon rock throughout Earth's history), according to Hutton, does not show a uniform sequential development over time, but rather a successive alternation between periods of deposition and times of violent upheaval. (We now understand unconformities to be erosional surfaces that separate younger from older layers of rock.) Believing in "a system of beautiful economy in the works of Nature," Hutton suggested that the same processes at work today were always at work slowly shaping the land.[8]

"FIGURED STONES" AND PALEONTOLOGY

Since ancient times, rock collectors have spotted *Lapides Figurati* (figured stones), strange images in rock that we now know as fossils. Although the Greek philosopher Herodotus fully understood what they represented, in the Middle Ages fossils were commonly thought of as an occult force at work within Nature, some direct evil force sent to mislead humanity, or even as radiated energy from the stars and planets.[9] By the beginning of the eighteenth century, opinion was divided as to whether figured stones were long-dead living things, or the result of strange forces beneath the Earth's surface. But in 1726 an illustrated book on fossils appeared that unwittingly brought the issue to a head. The book was called *Lithographiae Wirceburgensis*, and it both made its author's reputation and destroyed him.[10]

The treatise was written by a University of Würzburg geologist named Johann Beringer. An avid collector of the fossils that were common in his area, he believed that fossils were the stone remains of once-living forms. Some of his students decided to "help" him in his search by preparing artificial fossils molded out of clay, baking them, and then scattering them around as practical jokes. The stakes rose with each field trip; he "found" bees in their hives or sucking honey from plants, all perfectly preserved in stone. Beringer continued to believe in these finds, even when they became clearly artificial—engravings of the Sun and Moon, and then Hebrew letters. Undaunted by even these telltale signs, Beringer continued his fieldwork until one day he picked up a specimen containing the Hebrew version of his own name.

Horrified by the cruel trick his students had played, Beringer began a second project, to locate and buy back every copy of his work. He spent much of the rest of his life doing this. But the final irony did not occur until after his death. Trying to raise funds to pay debts, his estate published a second edition of his book!

Thanks partly to the *reductio ad absurdum* nature of Beringer's work, most scientists accepted that fossils were certainly the remains of once-living things.[11] In fact, the notion that the highest mountains contained fossils of sea animals encouraged catastrophist thinking—some massive event must have happened to thrust ocean-floor sediments to mountaintop heights.

In 1749 Buffon published his *Théorie de la Terre*. He went back to the earliest days of the Earth, in the wake of a collision, he thought, between a comet and the Sun. As a result of the crash, a small piece of the Sun detached and cooled to become the Earth. The "first form" of the Earth, Buffon believed, was filled with events of shock, agitation, and upheaval. Although the mechanisms of Nature in the early Earth were the same as they are today, their energy was far greater, in Buffon's view, so that changes happened under more active power and higher temperature.[12] Buffon's ideas had an influence on Georges Cuvier, perhaps the best-known catastrophist. Cuvier's ideas of revolutionary changes in the history of the Earth coincided with the tumult of the French Revolution. Born in the principality of Montbéliard in 1769, Cuvier grew up in an agricultural environment that was poor but emphasized learning. The town was almost a single fortress, and it was governed in a strange fatherly way. Each year the town mayor met with the village elders and the pastor to investigate family disputes—the town gave incredible pressure toward self control.[13] Although Cuvier was named Dagobert, he was addressed by the name of his older brother, Georges, who had died the year he was born. Cuvier became interested in the history of the Earth at age twelve, when his uncle gave him a copy of Buffon's work. Cuvier never had much interest in any natural event that could not be confirmed by any observational evidence, so he did not speculate much about the formation, or cosmogony, of the Earth. Instead, he concentrated on the events after our world was already formed. He believed that the fossil record, with its sudden delineations where rock types change and their included fossilized life forms, was evidence of catastrophic events. In order for these changes to be as rapid as they appeared in the rocks, the "energy" that caused them must also have been great, in fact far greater than any contemporary episode could handle: "no cause acting slowly could have produced sudden effects."[14] When Cuvier was twenty, his country was torn asunder by the storming of the Bastille and the onset of the French Revolution. It was a time of incredible violence, and it may have had an influence on the young scientist's thinking. To Cuvier, it seems that the violence of the French Revolution mirrored violent catastrophes in the Earth's past.

Cuvier had observational evidence for his belief. He had a special interest in the Cretaceous-Tertiary boundary, which is the site of a major extinction, and in the boundary between the more recent Tertiary and Quaternary periods. He established that certain life forms were confined to particular sedimentary rock layers, and that they were extinct above these layers. In 1812 he published his greatest work, *Recherches sur les Ossemens Fossiles.*[15] In it he proposed that the history of life on Earth has been marked by sudden catastrophes. In his model, the processes we now observe on Earth, like weathering and erosion, and mountain building, are not sufficient to explain the geological events of primordial times. Other mechanisms, absent today, caused whole continents to rise almost instantaneously, and major floods to cover huge areas of land. Unlike Buffon, Cuvier felt that whatever caused these events was not just an event of greater energy than is seen today, but of a completely different kind as well. "It is in vain," he wrote, "that we search among the powers that now act at the surface of the earth for causes sufficient to produce the revolutions and the catastrophes. . . ."[16] It is important to make this distinction, for even though he was unaware of them, we will see that cosmic impacts—which we now have witnessed with the collision of Comet Shoemaker-Levy 9 with Jupiter—qualify as such a cause. Although Cuvier believed that extraordinary episodes took place to produce the Earth's major changes, he did acknowledge the slowly evolving uniformitarian sequence of events that must have taken place between the violent outbursts of geologic activity called catastrophes.

By 1830 the geologic stage crossed over to England, where Charles Lyell published his *Principles of Geology*. The book's subtitle explained its aim: "being an Attempt to explain the former changes of the Earth's surface by reference to causes now in operation."[17] It was an utter repudiation of the catastrophist school and ushered in a century and a half during which the teaching of uniformitarianism was standard. Its frontispiece was an image of the Roman pillars of Pozzuoli, which Lyell considered a symbol of uniformitarianism, since they contained fossil marine animals that indicated gradual changes in sea level during historic times. However the pillars show evidence that is inconsistent with uniformitarianism; lower in the masonry, the numbers of marine animals are dramatically lower. Could this monument to uniformitarianism also contain an indication of a catastrophic change in sea level?[18]

Even in the nineteenth century, some scientists held open the possibility of major catastrophic events. Joseph Prestwich, for example, thought of a combined uniformitarian-catastrophist view, in which the *kind* of events never changed throughout geologic time, though the degree did. Between the time of Lyell and the mid–twentieth century, uniformitarianism became so firmly in control of geologic thought that catastrophism was treated as almost a historical joke. "If we must choose between these two concepts," wrote one author in a college text in geology published in 1971, "then uniformitarianism is the obvious choice. The very word 'catastrophism' has a bad connotation for most earth scientists. They are so adamant in refuting the more extreme views of the catastrophists that they prefer not to use the word at all."[19]

Plate tectonics was one strong reason for this belief. In the 1920s, Alfred Wegener, a German meteorologist, made the connection that the matching continental shelf lines indicate that the continents might have all been at one time a single continent. He called this ancient place Pangaea, for "all the lands."[20] In the late 1960s, a new field of study called paleomagnetics proved this idea and opened the field of plate tectonics. Maps of magnetic pole reversals show that oceans like the Atlantic and Indian are getting wider due to new midoceanic ridges that allow rocks from the mantle to rise and spread, and that other oceans are shrinking as the ocean crust dives into the Earth in deep ocean trenches. The vehicle driving all this motion is a series of continental plates that move about the Earth.

Plate tectonics was a uniformitarian revolution. Geologists specializing on one continent could now be worldly, because of the probability that formations on one continent were laid down at the same time and under the same circumstances as formations on another continent thousands of miles away. Because the continents separated and moved apart gradually, plate tectonics seemed to be a triumph of the uniformitarian view of looking at the Earth. "We were overwhelmed," notes geologist Walter Alvarez, who would later help bring down strict uniformitarianism. "Oceans so wide that they take hours to cross in jet aircraft have grown by seafloor spreading over tens of millions of years, at the rate of a few centimeters per year—about the rate at which your fingernails grow. Plate tectonics was the most gradual, uniformitarian theory imaginable."[21]

In 1970, geologist Richard Hooykaas wrote in a small monograph that catastrophism was as well founded philosophically as any other line of reasoning, and concluded that in the long run, the long-running war between uniformitarianism and catastrophism was a disagreement on method. Uniformitarians held that data had to be interpreted on the basis that present and past causes were similar in kind and degree, while catastrophists believed that geologic facts force an adaptation in our reasoning when rare, vastly larger forces occasionally happen and dominate what we see in the geologic record.[22]

If a philosophical principle called Occam's razor is applied, then uniformitarianism is the most reasonable way to explain the vast procession of life forms on this planet. The "Razor" was first developed by the fourteenth-century Franciscan William of Occam, who suggested that where two or more explanations fit the observed facts, the simplest one is probably the correct one: "What can be done with fewer is done in vain with more."[23] If we are to accept Occam's razor, and if uniformitarianism's slow and known processes are pitted against the unknown and rapid effects of a catastrophe, the processes that uniformitarianism demands must be the way the Earth evolved.

But applying Occam's razor does not necessarily render a philosophy right. Is it possible that Lyell's *Principles of Geology* had everything written up, as Gene Shoemaker would later think, a bit too neatly? Were the great extinctions of life, like the one at the end of Earth's Cretaceous period, the result of gradual change? Like many modern scientists, paleontologist Stephen Jay Gould, one of the major thinkers of our time, has questioned whether the history of Earth has not always followed the creed that uniformitarianism set for it. In 1962 Thomas Kuhn suggested that the problem may be more serious than a succession of scientific theories: It may be in the conduct of the scientific method itself. Should that method be gradualistic and uniformitarian, a slow and steady search for truth? Kuhn saw science as a political structure where paradigms last, wiping out anomalies, which are ignored or dismissed. Occasionally new paradigms can replace old ones.[24] In the 1970s Gould's theory of "punctuated equilibrium," also known as punctuational change, proposed that occasional sudden change dominates the history of life on Earth, and that relatively rapid flips took place between long periods of stable equilibria.[25] Beginning with sudden onset of rapid speciation, the clock then enters a long period of little change before the species either becomes extinct or survives the next cataclysm. "In my own work," he writes, "I have been impressed by the powerful and unfortunate influence that gradualism has exerted on paleontology via the old motto *natura non facit saltum* ('nature does not make leaps'). Gradualism, the idea that all change must be smooth, slow, and steady, was never read from the rocks. It represented a common cultural bias, in part a response of nineteenth century liberalism to a world in revolution." Gradualism, Gould argued, was imposed on geology as a "bias upon nature:"

> Nature had an input in the formulation of gradualism—some of her processes work slowly and cumulatively. But the doctrine of Lyell's third uniformity has more complex roots. I am convinced that the cultural and political context of European society had an input equal to, or greater than, Nature herself. The notion that science operates apart from culture by a universal method that yields truth according to canons of observation and experimentation is a myth that has been carefully nurtured by scientists themselves. Science operates, as does all creative thought, within a cultural context that influences all practicioners in various subtle and unacknowledged ways.[26]

The rise of this new catastrophism almost two centuries after Sir Charles Lyell thought he had rid geology of this kind of thinking is a most unusual event in the history of scientific thought. For many years geologists thought of their science's catastrophist past as being not much better than that of ancient alchemists who would turn any substance into diamonds, or the philosophers' dream of answering the question of how many angels could dance atop the head of a pin. But in the case of geology, those who ascribed sudden changes to events akin to Noah's flood may have had a point; in fact, a kilometers-high tsunami is an inevitable result of a major ocean impact by an asteroid or a comet.

7. Gene and Carolyn at the eighteen-inch telescope. They are standing next to Dr. George Stevens's poster presentation suggesting a possible impact site on the South Mountain Batholith in Nova Scotia. Jean Mueller photo.

Does this new thinking make uniformitarianism obsolete? Hardly. Now that we have seen an actual impact, the 1994 collision of Comet Shoemaker-Levy 9 with Jupiter, we can certainly add comet and asteroid impacts to the chain of events that can be expected to occur on Earth. Impacts have been and are a part of the normal chain of events on Earth, said geologist George Stevens of Acadia University—who discovered a 500-meter-wide impact site in Nova Scotia—so they are a part of Earth's uniformitarian past.[27] Gene Shoemaker would have agreed, as he and colleague George Wetherill wrote in 1982:

> Although the physical encounter with the Earth of these objects can properly be termed "catastrophic," in terms of the magnitude of the effects they produce, they are at the same time "uniformitarian" in that they represent the extension of presently observed geologic

processes to earlier geologic time. The preserved record of cratering on the Moon shows that prior to 3.8 billion years ago, the rate of impact of residual pre-planetary bodies with the primitive Earth must have been very much higher than the present rate. During subsequent Archean, Proterozoic, and Phanerozoic times, impact of extraterrestrial bodies up to tens of kilometers in diameter should be thought of as ordinary and inevitable phenomena characteristic of the well-behaved modern solar system.[28]

In the nineteenth century, both schools of thought were affected by the swirl of political events around them. Lyell's uniformitarianism was as affected by the culture of England as Cuvier's catastrophism was motivated in part by the violence of the revolution in France. After more than a century and a half of uniformitarianism, the geological stage was set for another look at a neocatastrophist philosophy. This time it would be based on solid evidence that the Earth has been a victim of events that would have stunned the imaginations of the great geologists of the past, and an event that would propel Eugene M. Shoemaker to the center stage of modern geological thought.

CHAPTER 6

Impact!

Turning and turning in the widening gyre

The falcon cannot hear the falconer;

Things fall apart; the center cannot hold;

Mere anarchy is loosed upon the world. . . .

—WILLIAM BUTLER YEATS, *1921*

APRIL 23, 1965, was the day I joined a group of friends to attend a lecture about the Moon, given by Donald McRae, one of Canada's most famous astronomers. I had no idea that the evening would open my mind to the possibility that cosmic impacts have affected the history of the Moon and the Earth. Some thirty of us had traveled by bus for two hours to Ottawa, Canada's capital city, to hear a review of the theories that explain the surface of the Moon. It was a good time to hear such a lecture, for the mid-1960s were a time of uncertainty about the nature of Earth's satellite. With humanity about to set forth upon this new world, geologists and astronomers were divided as to whether the lunar surface was a result of volcanic forces from beneath, or cosmic forces from above. Of course, we were treated to other explanations of the lunar surface as well, like green cheese, reindeer moss, fairy castles, and cotton candy. The green cheese theory, it turns out, came from the satirical mind of Erasmus of Rotterdam, who in 1508 suggested that the Moon resembled green cheese, referring to new and immature cheese, round and with a mottled surface.[1] By 1965 these whimsies had been discarded in favor of a more scientific approach. In that year some thought that the lunar surface, including most of its craters, was formed from volcanic activity from beneath the surface. It was time to look back at an old theory, originally proposed by Grove Carl Gilbert in 1893. That famous geologist suggested that large asteroidal objects impacted the Moon, forming its craters.[2] In 1949, Ralph Baldwin had articulated that the Moon's craters were mostly of impact origin[3] and Gene Shoemaker revived the idea again around 1960. Since the time he had used his surveying telescope to look at the Moon from the clear sky of the Colorado plateau during the late 1940s and 1950s, Gene, almost alone among geologists, saw the Moon as a fertile subject for field geology. He saw the craters on the Moon as logical impact sites that were formed not gradually, in eons, but explosively, in seconds.

Three months after that lecture in 1965, the American *Mariner 4* spacecraft sped by Mars. On July 15 the craft sent back pictures of a world apparently covered with craters. Were the wasted surfaces of Mars and the Moon what most planets were like? With the discovery of Mars's crater-scarred surface, impacts forced themselves to the forefront of geological interpretation as a viable means of shaping the surfaces of worlds.

EARLY IDEAS ON IMPACT CATASTROPHE

Even in the nineteenth century, one astronomer paid some attention to the possibility of an impact on Earth, perhaps by a comet. In the late 1850s, J. Russell Hind predicted that the bright comet seen three centuries earlier might return. Called the comet of Charles V of the Holy Roman Empire, the comet was thought to have a cycle of about three hundred years. In his thoughts about the comet, Hind wondered what would happen if it collided with the Earth. "At worst," he wrote, "a direct collision would perhaps be comparable only as regards the mechanical effect upon the earth to a meeting with a huge cushion."[4]

The seed of an idea that sudden extraterrestrial causes could have had massive and sudden effects on the pageant of life on Earth was planted in the fertile mind of Georges Cuvier. Even though some of his reasoning was based on his belief that geological events should be tied into the Book of Genesis, his idea that brief catastrophic intervals occurred from time to time in Earth's history was scientifically correct. Toward the middle of the nineteenth century, Alcide d'Orbingny listed no fewer than twenty-seven major extinctions, or "revolutions," in which life on Earth was substantially changed. These bio-events became the foundations that geologists during most of the nineteenth century used to subdivide the Phanerozoic, the eon, or span of geologic time, that includes all evidence of past life on Earth. The geologist John Phillips specified that the major divisions of life be divided into the three eras of Paleozoic (ancient life), Mesozoic (middle), and Cenozoic (recent). However, even as late as the 1960s, some geologists doubted that some mass extinctions, where a substantial percentage of all species of life are extinguished, were synchronously global in nature.

Discoveries in astronomy were also pointing the way toward the change of thinking about Earth. The first asteroid was found on January 1, 1801. It and many others discovered subsequently seemed to orbit randomly between the orbits of Mars and Jupiter. In 1866 Daniel Kirkwood discovered that Jupiter affects the orbits of asteroids, resulting in differences in how asteroid orbits are arranged. In 1918 astronomer Kiyotsugu Hirayama suggested that some asteroids belong to families of asteroids in similar orbits, derived from the breakup of a parent body by collision with another asteroid. In 1930 Karl Reinmuth discovered Apollo, an asteroid actually crossing the orbit of the Earth. These discoveries were not about catastrophes on Earth, but were steadily pointing the way to providing a large army of asteroids capable of causing such catastrophes.

Comets were showing that they can also pack a punch. In June 1770 a comet called Lexell tore by the Earth at a distance of only 1.4 million miles, just 175 Earth-diameters, a very small distance by cosmic standards, and some ninety years later the Earth passed through the tail of the Great Comet of 1861. In 1910 the Earth actually passed directly through the tail of Halley's Comet. Later in the century Comet IRAS-Araki-Alcock passed by the Earth at a distance of less than 3 million miles, followed by Hyakutake in 1996, at a distance of 9 million miles.

In 1950 *Worlds in Collision* was published, a book that became very popular. Its author, Immanuel Velikovsky, inferred from historical facts and mythology that an unlikely series of comet impacts caused some biblical events, like the Egyptian plagues and the parting of the Red Sea.[5] The details of Velikovsky's ideas were so strongly ridiculed that his theme that impacts have indeed affected the Earth, was clouded and lost.

In 1954 the German paleontologist Otto Schindewolf proposed that mass extinctions, particularly the one at the end of the Permian period 225 million years ago, were the result of cosmic radiation caused by supernova explosions.[6] This was a remarkable idea, a true leap of faith for a highly regarded scientist. A supernova marks the end of the life of a massive star: as its core collapses, the star explodes. For a few weeks or months the supernova emits radiation that is equivalent to the intensity of a hundred billion suns. That cosmic radiation, Schindewolf thought, could be lethal to life on Earth if the exploding star were close enough to Earth. He proposed that the radiation from a supernova would have two effects. One is the extinguishing of many species of life. The second is more problematic, to say the least: the supernova's radiation could cause macromutations that would produce healthy, and instant, successors to the dying species! Schindewolf was ridiculed: how could a new family begin with a single accident, like a bird coming forth from the egg of a reptile?[7] Although Schindewolf himself was not comfortable with this idea, he could not explain the sudden changes in the fossil record in any other way.

As intriguing as the supernova theory may have been, it did not receive much attention at the time because there was no confirming evidence. In 1956 paleontologist M. W. de Laubenfels proposed a variation of the cosmic cause theory by suggesting that the dinosaurs died from heat associated with the impact of a large meteorite.[8] The paper appeared under the somewhat embarrassed title "Dinosaur Extinction—One More Hypothesis," almost as if its author didn't want anyone to consider it seriously! In 1963 Schindewolf published another paper with the word *neocatastrophism* in its title, and so he is credited with the coining of the term used today to define a rebirth, in a different form, of the new school of catastrophist thinking.[9] A decade later Harold C. Urey suggested that sudden extinctions, like those of the dinosaurs 65 million years ago, were caused by comet collisions. He predicted that microscopic versions of glassy tektites, blobs of molten rock made from rocks ejected from terrestrial impacts that melted rapidly and then cooled quickly as they fell again to the ground, might be found at the boundary between the Cretaceous and Tertiary periods.[10]

WHERE'S THE BEEF?

Two decades after Schindewolf's supernova proposal, little had changed to suggest that cosmic events could cause massive changes in Earth. Victor Hughes of Queen's University discovered two pulsars 456 and 196 light years away that might have affected Earth in a small degree 62,000 years ago and 440,000 years ago.[11] However, the remains of a supernova from 65 million years ago would be extremely hard to locate, let alone date. Evidence for a cosmic cause was not strong until Luis and Walter Alvarez found it, in the 1970s, amid the beds of a pinkish limestone called Scaglia rossa. As the Alvarez team worked for several summers on these beds, they became intrigued with a boundary layer of clay, one centimeter thick, separating major eras of the Mesozoic (middle life) and the Cenozoic (recent life) on Earth. As the lime muds were being deposited peacefully underwater, something catastrophic must have been happening above water. Although the same kind of rock continued to form, there was a major change in the life forms that were fossilized in it. In the Cretaceous beds there were fossil shells of numerous species of foraminifera, one-celled marine life that lived out their lives in the sea. The foraminifera were as large as grains of sand and very numerous in the uppermost limestone beds of the Cretaceous. In the younger limestone above, separated by the clay layer, there were far fewer foraminifera fossils, and those that did exist were much smaller and lacked most of the species that had existed earlier.

Most species of these tiny creatures became extinct at the same time as did the huge dinosaurs and many other species of life. The Alvarez team decided to examine the nature of the thin boundary clay itself. Whatever happened at the boundary seemed to have taken place in a period lasting no longer than 100,000 years. For an extinction this large to take place this quickly, the normal uniformitarian processes could not have been responsible. In June 1977 the team found something extraordinary in the boundary clay, an anomalously high level of iridium that seemed to have been deposited in a very short period of time, perhaps as short as a few months. In the rocks below the boundary clay, thousands of foraminifera were present in each cubic foot of rock. Above the clay the foraminifera were virtually gone. The iridium layer indicated that this major mass extinction event was too short to have resulted from a prolonged period of volcanic activity. Moreover, iridium itself pointed to a cosmic source: almost all of the cosmically abundant material is absent in the Earth's crust, having accompanied iron into the Earth's core when the core formed.

Early in 1980 Kenneth Hsü published a paper in *Nature* suggesting that late Cretaceous extinctions were caused by a comet impact.[12] But the Alvarez paper in *Science*, with its powerful evidence of iridium from a cosmic impact, was a central development in the geologic literature.[13] The Alvarez work offered an explanation for the mass extinction at the end of the Cretaceous that was decidedly nonuniformitarian, and it opened up a decade of controversy. If there was an impact, where was the crater, asked geologists Charles Officer and Charles Drake in the mid-1980s. They proposed that the iridium could have come from a hundred-thousand-year period of heavy volcanic activity,[14] a theory disputed by impact scientists as well as some volcano-geochemists.

By the end of the 1980s, some one hundred sites with significant iridium anomalies had been found worldwide. Some outcrops contained small, strangely deformed grains of quartz. To deform quartz requires a shock greater than even a volcanic steam explosion like the Krakatoa one in 1882 could provide. At one of these sites along the Brazos River in Texas, astrogeologists Alan Hildebrand and David Kring found a variety of rock fragments that showed evidence of rapid deposition by a tsunami several hundred meters high, moving hundreds of kilometers inland. They noted that these deposits, which were several centimeters thick, cluster around the Caribbean basin. Could a tsunami have roared northward from the Gulf of Mexico? In addition to the iridium and shocked quartz, Hildebrand and Kring found tektites even larger than Urey had predicted. And in the small nation of Belize, out-croppings of boundary layer were found that are far more than a thin clay layer; these examples are a hundred feet thick, filled with many types of boulders that could have been hurled out of an impact site. The two scientists concluded that the best candidate site was a buried area centered in Mexico's Yucatán peninsula.[15]

Just one year after the iridium announcement, Glenn Penfield and Antonio Camargo, geologists working for Mexico's Petróleos Mexicanos oil company, or Pemex, reported at an oil exploration geology meeting that a hundred-mile-wide feature on the Yucatán previously thought to be a buried volcano, might actually be the remains of a gigantic impact crater. Their work was initially over-looked by Alvarez and virtually everyone else until 1991, when Hildebrand and Kring focused on this site. A series of rocky cores drilled out of the deep crater by Pemex geologists added weight to the idea this was an impact crater. In 1991 David Kring first de-tected the telltale evidence of shocked quartz in the crater, the best evidence yet that the feature was the result of an impact.[16]

THE RECORD GROWS

Geologists were now focusing on other major episodes of mass extinction. The earliest known evidence for a massive change in life is at the boundary between the rocks of the Precambrian and the Cambrian, more than half a billion years old. That boundary is marked with a sudden proliferation of life forms. After a billion years of only one-celled life forms, there was an incredible burst of speciation that included creatures with mineralized skeletons, like trilobites. Also appearing during this period were chordates, the earliest members of the phylum that would eventually include hu-mans. Could a sudden cosmic cause like one or more impacts have started this diversification? There is a small layer of iridium at that boundary, too, and a 55-mile wide impact structure at Lake Acraman, in South Australia, that could be dated to that time. But 500 million years is a very long time ago, and the evidence is spotty.

An extinction at the end of the Cambrian might have been the result of a worldwide rise in sea level. About 450 million years ago, the close of the Ordovician period was punctuated by the oldest known mass extinction. Moreover, the formation and erosion of the Gondwanan ice sheet, a long period of glaciation that would not normally be called for after an impact, accompanied this loss of life.[17]

A second major loss of species of life, including reef-building corals, took place during the Devonian period. The discovery of impact-formed glass spherules in China and Belgium adds evidence that this loss of species was related to at least one impact. The late Devonian (see table 1) is a fascinating period of geologic time, during which the first forests appeared. The fernlike trees would have been an alien scene on our own planet. More important, the dominant form of life at the time, the arthropods, left the waters to become land dwellers. Is it possible that the first major inhabitants of Earth's lands were greeted by the unfortunate but incredible show of a major impact?[18]

Some 250 million years ago the end of the Permian period was punctuated by the third and most dramatic mass extinction in the recorded history of the planet. Fifty-seven percent of the families, or a whopping 96 percent of all the species of life were extinguished at that time. The fact that geologists are looking for an impact cause for this episode shows how far geology has progressed in only three decades. In the late 1960s, geological thought centered on the new concept of plate tectonics. The continents rested on larger crustal plates, and these plates were shifting across the globe, carrying the continents with them. During the Permian, the Earth was a very busy place; moving tectonic plates were rafting the continents of Gondwanaland and Laurasia together to form a single supercontinent we call Pangaea. As sea level fell, parts of the ocean basins were deepening. Could the worldwide massive volcanic eruptions associated with this activity have led to the mass extinction? Recent isotope studies suggest that a gradual decline of oxygen levels might have been the cause.[19] At the end of the

Permian period, Siberia, then a separate continent, "collided" with the eastern edge of Pangaea, accreting to the Pangaean land mass that then comprised virtually all the land on Earth. This union was accompanied by the formation of the Siberian traps, where thousands of cubic kilometers of lava spread out over some five hundred thousand years. These eruptions affected Earth's climate, which could have contributed to the great extinction.[20] However, recently there has been a finding of two widely spaced iridium anomalies, and in the south Atlantic ocean on the Falkland Plateau, are two structures, one of which is two hundred miles wide, that could be the remains of craters formed from impacts at the end of the Permian. Some geologists suggest that this extinction occurred over a period no longer than fifty thousand years, instead of millions of years.[21]

More than two hundred million years ago, the Triassic, the first period of the Mesozoic era and the time of the dinosaurs, ended with a fourth mass extinction that is under increasing suspicion of being related to a series of impacts. Quebec's Lake Manicouagan, a sixty-mile-wide crater in the northeastern part of the province, was formed at about this time, and four other major craters may be related to it. Three craters—Rochechouart in France, Manicouagan in Quebec, and Saint Martin in Manitoba—were at the same latitude at the time of the end of the Triassic. The other two, Obolon in Ukraine and Red Wing in Minnesota, lay on paths with Rochechouart and Saint Martin and could have been secondary effects from those impacts. Their paleoalignment, or alignment at the time of their formation, was pointed out in 1998 by a team of scientists centered at the University of Chicago's Paleographic Atlas project.[22]

A fifth major extinction took place some 91 million years ago, in the middle of the Cretaceous period some 26 million years before impact six, the big event at the end of the Cretaceous. Its cause may have been an extended period of volcanic activity. About 26 million years after this event, at the end of the Eocene, a seventh mass extinction took place, but in steps. Most recently, a major extinction of species, in the late Pleistocene, is credited to postglacial warming and predation by early man.

TABLE 1
The Geological Time Scale

Era	Period	Time began (in years)
Cenozoic	Quaternary	2 million
	Tertiary	65 million
Mesozoic	Cretaceous	136 million
	Jurassic	205 million
	Triassic	225 million
Paleozoic	Permian	280 million
	Carboniferous	315 million
	Devonian	345 million
	Silurian	395 million
	Ordovician	430 million
	Cambrian	540 million
Precambrian	Proterozoic	2.5 billion
	Archean	4.6 billion

In 1984 D. Raup and J. J. Sepkoski suggested that extinctions take place on our planet every 26 million years on average. Other researchers have confirmed that some periodicity seems to exist. This cycle may be genuine, although it is controversial.[23] It does coincide roughly with a sine wave-type motion of our solar system as it races through the dust-laden plane of our galaxy once every 26 million years. As it does so, the mass of these giant molecular clouds might trigger more comets to head into the inner solar system. If this cycle is correct, then it allows us to achieve, at last, a neatly wrapped consensus between the uniformitarian and neo-catastrophist schools. In the evolution of our planet, its regular passages through the dust-rich plane of the Milky Way galaxy result in a greater chance for catastrophic events every 26 million years. Perhaps a cycle of impacts is indeed part of the uniformitarian evolution of Earth, just as impacts are an inherently uniformitarian consequence of planetary formation and evolution.

The whole question of how large-body impacts had affected life on Earth was, by 1983, one of Gene Shoemaker's central interests. In that year Luis Alvarez and David Raup invited a handful of scientists to attend an intensive two-day workshop during which two major ideas about impact periodicity were debated.[24] One was the passage of the solar system through the galactic plane, and the other was the existence of a companion to the Sun, whose 28-million-year orbit would periodically upset the orbits of distant comets, hurtling them into the inner part of the solar system where they would threaten the planets. Gene did not support the theory that the Sun has such a "Nemesis" companion star.

However they affect geologic thought, the impacts of comets and asteroids have added a bold new dimension to our understanding of life on this planet. Whatever school of historic reasoning we put them in, learning about their importance has made our Earth a much more exciting place.

CHAPTER 7

A Shot in the Dark: 1953–1960

How soon hath Time, the subtle thief of youth,

Stolen on his wing my three and twentieth year!

My hasting days fly on with full career,

But my late spring no bud or blossom show'th.

—JOHN MILTON, *1645*

GENE'S EARLY CAREER was creative, and filled with work in the field. His future work, the uncovering of the role and nature of impacts, stretched out before him. Gene and Carolyn were young, the dream of the Moon had not become a race, and field research out in the countryside was still unencumbered by fame and administrative responsibilities.

Early in 1953, the Shoemaker's first child, Christy, was born. The baby's first months were spent on the Colorado Plateau in Grand Junction, but with the arrival of summer it was time to head, at last, back to Princeton so that Gene could complete his second and final year of course work. But their Ford, Jezebel, pulling a trailer carrying the Shoemaker family and its possessions, overheated on a flat, straight Nebraska highway. Gene had an idea. Propping the hood partway up so that more air would reach the engine, he continued driving. The traffic was light, but the young scientist had to peer awkwardly out his window, wind and bugs flying into his face, as he drove down the road. Although he did slow down a bit, the ploy worked and Jezebel made it through Nebraska without further incident. Once they got to Princeton, the family resided in a small house in the country, a short distance from Princeton. Gene worked at Princeton with Harry Hess, a geologist

8. The youthful Gene enjoyed wearing an engineer's cap.

completely in tune with his interests in the Moon; in fact, Hess would later organize a committee to determine who could be a scientist-astronaut.

The following spring, Jezebel was retired at last, and the Shoe-makers returned to their second home, in Grand Junction, in Sweet Sue, a new black station wagon. Bringing his family and friends as much as possible into his own work seemed a natural goal to a man who was overwhelmed by the magnificent story of planet Earth. But in the rugged environment of a geologic field study, it

9. Gene, with Christy, Linda, and Pat.

was not (and still isn't) customary for spouses to join in field trips. Gene never could understand why, and he encouraged Carolyn to join him in the field as much as possible. Gene also saw no reason why children should not accompany their parents in the field. For the young Shoemakers, geological field work was a part of their lives. The couple made a point of bringing their baby daughter along. As Christy grew, she soon was accompanied by brother Patrick, and sister Linda, and the family of five often did fieldwork as a group.

One of their field trips was an effort to map a diatreme in the Hopi Buttes. With two other geologists, Carl Roach and Frank Byers, Gene and Carolyn set up their "stationary camp," a camp where they planned to stay for the duration of the mapping. Nearby was an old water tank out of which stretched a hose. Its purpose was to provide a convenient way for a user to fill a personal tank with water from the hose and then leave a quarter for the service. On the first day of camp, Gene took water collection duty. With his containers in tow, he drove over to the tank and surveyed the water supply with some puzzlement. The hose was not nearly long enough to reach his tank! Without any spare parts or hoses, Gene shook his head in frustration and tried to figure out a solution. He carefully tried to position his tank under the short hose, and then holding it in place with one arm, he reached over, somewhat acrobatically, to the knob with the other. He turned the knob as gently as he could, but a torrent of water gushed out, soaking Gene and his tank. Eventually he managed to fill the tank. He dried off and returned to camp. On the way back to camp Gene bristled with frustration over how poorly set up the water supply was. But as he drove, his expression changed to a grin as he decided to turn the event into a practical joke. When Gene returned to camp he told Carolyn what had happened, but no one else. The following week, the completely unsuspecting Roach went to fetch the water and returned soaking wet but with another week's water supply. He returned, had a good laugh with Gene and Carolyn, and waited till the third week. By this time there wasn't much left of the apparatus at all, and Byers returned to camp utterly drenched and fully aware that he'd been had![1]

The diatreme studies that Gene did on the Hopi Buttes would later prove useful in planning for the Apollo missions. Gene thought that it was a good idea to look for a lunar site, like the Davy catena, that might be a structurally controlled string of volcanic craters. If that was true, then the astronauts might find xenoliths, or rock fragments blown out from beneath the lunar surface, that might show the deeper structure of the Moon.

Gene was intensely enthusiastic in those early years. "He couldn't visit anyone without telling them of his enthusiasm for whatever he was working on," Carolyn recalls. The family would often visit the nearby Indian trading post at Indian Wells. As one Thanksgiving approached, the post's owner, a man known as Hukriede, invited them for a dinner complete with turkey, ham, and three kinds of stuffing. This event was special and fun, even though Gene did not easily relate to his young children. "Although family was important to Gene," Carolyn says, "he was not a baby person or a small child person. He wanted someone he could talk to. I'd seen him joking and playing with young kids. And he did that with ours, but sometimes his work absorbed him more than his young children."

A ROAD TO THE MOON

Gene Shoemaker's search for uranium seemed to be leading to somewhere in a northern Arizona field of old volcanoes called the Hopi Buttes. His earliest major discoveries as a geologist, in fact, were deposits of uranium in the eroded volcanic vents of those long-extinct volcanoes. But it was the ancient volcanism itself that really captured Gene's attention. The surrounding craters seemed similar to what he had seen in pictures of the Moon. Was it possible, Gene thought, that lunar craters were the result of a similar process?

Two years after his first twilight visit to Meteor Crater, after he read a paper by geologist Dorsey Hager, Gene's attention turned toward it again. This new research focused attention on small structures called evaporites that oil diggers had dug up a few dozen miles east of the crater. Evaporites are the signature of salt flats, and they led Hager to suggest that the crater was the result of one that had collapsed. Although Gene doubted that theory, he was intrigued by the paper and decided to return to the crater and try to test Grove Karl Gilbert's conclusion from several decades earlier that the crater was the remains of a volcanic steam explosion. Hager sent Gene a sample of pumiceous silica glass that they thought could be either volcanic or melted sandstone. Gene arranged for spectrographic analysis. The result was stunning: the glass was Coconino sandstone, the same rock that is present in such quantities in the Grand Canyon, but its quartz was fused somehow, the end result of a process that involves temperatures of about fifteen hundred degrees Celsius, some three hundred degrees higher than the hottest lava flow. No, the crater could not be the result of volcanism, nor could the dynamics of a collapsing salt dome explain temperatures as hot as this. Gene began to suspect that the only mechanism that could generate this much heat was the explosive impact of an asteroid from space as it struck the Earth at incredible velocity.

The Mouse That Roared

Despite these tantalizing clues leading him toward Meteor Crater, Gene decided to wait on his crater investigation and return to his uranium search on the Hopi Buttes. It was now 1956, and uranium, the magical source for the nation's fledgling nuclear power industry, was proving far more abundant than expected. The problem for geologists in 1956 was no longer uranium, but plutonium, the element first produced in 1940 by a team led by Glenn Seaborg and which formed the heart of the United States nuclear arms race. Gene now became part of a new program called Project MICE— for Megaton Ice-Contained Explosion. If successful, a blanket of uranium wrapped around a nuclear device would produce, in a one-megaton nuclear detonation, a supply of plutonium. To be successful, the test needed to occur in a cavern contained by ice. Since the United States had no sufficiently large bodies of ice under its control to work with, the search began for an alternative, in fairly pure bodies of salt. The question Gene needed to explore involved the nature of the underground nuclear blast: would it lead to a surface eruption, a modern remake of the ancient volcanoes of the Hopi Buttes?

In the course of his study, Gene visited the artificial craters called Jangle U and Teapot Ess. The result of a small, 1.2-kiloton nuclear explosion, Jangle U was formed instantly near the end of 1951, and Teapot Ess was carved by a similar explosion in the early spring of 1955. These craters are about one hundred meters across. Gene noticed that rocks that had been melted by shock forces were spread out along the floors of these craters, suggesting that whatever plutonium would be produced in a megaton explosion would not be easily accessible on the floor of the explosion cavity but would be dispersed through the broken rock.

But Gene noticed something in those two craters that interested him far more than plutonium—an uncanny resemblance to the form and structure of Meteor Crater on Coon Butte. Comparing the kiloton of energy that went into carving out each of these craters to the incredible amount of energy that went into the larger, natural crater led him to the near certainty that the Arizona crater was created almost instantaneously, by an impact. "I was astonished to discover that the structure of Meteor Crater was pretty much a scaled-up version of that of Teapot Ess," he remembered. The impact would have been enough to disintegrate the main body of the iron mass that fell there, leaving only a few tons of Canyon Diablo iron meteorite scattered across the landscape. These remaining pieces would have survived by breaking away from the main body seconds before impact. In one intuitive leap, Gene realized why there was so little meteorite left for Barringer to mine, and similarly why MICE wouldn't work to produce a supply of plutonium—it would be widely dispersed and essentially not gatherable. For Gene, the failure of Project MICE translated into a golden key that he would later use to unlock the door of a new science.

If the crater was really the result of some "small star," as Grove Karl Gilbert had put it almost a century ago, what kind of celestial body would have been responsible?[2] The night sky is filled with small meteors; these are events that occur when tiny particles of asteroids or comets encounter Earth's atmosphere. As they heat the surrounding air to incandescence, the air glows brightly for a second or two. Occasionally, a larger object the size of a marble might strike, lighting up in the night sky as a fireball. Football-sized objects also encounter the Earth; we see them as bright, often exploding bolides.

The larger an impacting object might be, the rarer is its chance to strike the Earth. In the 1950s the only place where the result of such a hit could be found, possibly, was at Meteor Crater. Gene set out to find the evidence that the crater was indeed of impact origin.

To intensify his search for evidence, Gene clearly needed more information in the form of other examples. Happily, Earth's nearest neighbor in space and the subject of Gene's growing dream, the Moon, has a large supply of craters. Since erosion on the Moon is negligible compared with that on Earth, once a crater is formed there from whatever cause, it lasts virtually forever. To answer his questions, Gene needed to study the Moon, and in 1956 he spoke with the Geological Survey director, Thomas Nolan, about the possibility of launching a photographic study of the Moon in order to produce the first geologic map ever made of a body other than the Earth. Thus Nolan became the second person after Carolyn to hear the details of Gene's dream to visit the Moon. Nolan didn't laugh. Instead, he steered the young geologist to those who had already thought of lunar topography. He soon learned that while a topographic map of the Moon had been thought of, a map that emphasized the remote world's geological structures had not. But this part of the Shoemaker dream would bide its time. In 1957 Gene's professional time was pretty much consumed by MICE. Besides, the spring and summer of 1957 was not the time to talk about mapping the Moon. Although the U.S. government was building and firing rockets at White Sands in New Mexico, the scientists' proposal to send a satellite into orbit as a prelude to scientific exploration that might include the Moon was soundly dismissed amid orders to not even discuss such ideas. Missiles were for the defense of the nation, they were told, and science was not their purpose.

In the spring of 1957 the Shoemakers left for Fort Stockton, Texas, and the Sierra Madre structure that lay nearby. Hollering out the words of *The Eyes of Texas Are upon You* as they crossed the border into the Lone Star state, the family set up camp with Richard Eggleton and his family on a windy ridge overlooking what Gene thought was an impact structure. A first-rate field geologist, Eggleton would later become a central figure in lunar science.[3] While their children slept in the van, Gene and Carolyn slept out under the stars—and under a comet. As the sky grew light, the comet rose to serenade the couple each morning before fieldwork began on the remains of what would prove to be a much more ancient visitor from space.

The comet was Mrkos, visiting Earth for the first time since its formation in the outer solar system more than four billion years ago. However, the earlier visitor had come to stay. Less than 100 million years ago it plummeted to Earth in a fiery crash and explosion. The winds and rains of time have eroded the impact site, so all that was left in 1957 were hills that marked the outer ring and the central uplift at the site.[4]

THE DAWN OF SPACE

October 4, 1957: In the cold light of dawn on a steppe deep in the Soviet Union, a rocket ignited and quickly built up thrust. As holddown clamps fell away, the rocket surged off its launchpad. Slowly at first, then picking up speed, the missile soared into the sky, carrying with it the first artificial moon ever placed into orbit around the Earth.

Sputnik, or *Fellow Traveler*, caught the imagination of the world and inspired the dread of defense planners throughout the free world. Ostensibly a Soviet contribution to a worldwide research program called the International Geophysical Year, *Sputnik* clearly raised the stakes in the Cold War to a startling new level. Questions were raised at once. Were Soviet scientists so far ahead of their American counterparts that the United States could not be competitive on the stage of space? Worse, was the Soviet Union planning a space platform from which to launch a nuclear attack on the United States?

As the world convulsed under this news, Gene returned from a MICE progress meeting back to camp in the Hopi Buttes and heard the news on a portable radio. Oh hell! he thought as he prepared to rejoin his camp in the middle of the lonely buttes, I'm not ready for that yet!

With *Sputnik*, American space policy changed radically. Instead of no satellite program, there was a frantic race to get the first satellite launched. The navy's research lab prepared its *Vanguard* satellite, while the army's *Jupiter C* was rushed into production with an *Explorer* satellite aboard. As Werner von Braun struggled to launch a satellite within ninety days of the *Sputnik* flight, *Vanguard I* collapsed back into its launchpad and exploded in flames. Early in 1958, a *Jupiter-C* rocket launched *Explorer I* into orbit. By the time *Vanguard 2* soared into orbit on February 1958, the entire nation was set to travel a new and exciting educational path, with science seemingly at the forefront. That year, under the guidance of Senator Lyndon Johnson, the National Aeronautics and Space Administration (NASA) was born. As Gene continued his MICE work, he began to think that a Moon program might come sooner than he had expected. In meetings with USGS senior staff he talked up the Moon, trying to arouse interest, but despite the rapidly developing space program the tradition-based Survey was still loath to take the Moon seriously.

But Gene did. His evolving interest was a turn away from his original thesis subject about the Colorado-Utah border salt structure, and it led to an uncomfortable situation back at Princeton, where time was running out for him to complete whatever dissertation he had in mind. As 1957 turned to 1958, Princeton's geology department noted in frequent reminders to Gene that his thesis was overdue. Gene had the best of intentions to write his thesis, but the demands of the Geological Survey and his own expanding interests kept him from finishing it promptly. Failure to complete his projects was a habit that would plague him throughout his career. His initial subject was the geology of the stunningly beautiful Sinbad and Fisher Valley salt structure that spans the border between Colorado and Utah. In the late 1950s, as Princeton's time limit was drawing close he gave a colloquium about his work on Meteor Crater. In the audience was his friend and thesis advisor, Harry Hess. "That was a good presentation," he told Gene. "It would even be a good subject for a thesis!" The possibility of switching his topic was tantalizing, and Hess worried that Gene's varied interests might mean he'd never finish a thesis on anything. Gene thought he'd better take the hint and get to work. He decided to work on the crater more intensely, and in the summer of 1959, submitted his Meteor Crater research to Princeton's Department of Geology.

Gene's renewed interest in completing his thesis had a sadder motivation. In February 1953 his father was injured in a fall. Standing twenty feet up on a platform with no rails, his rope broke. He fell over backward, doing a flip and landing on his feet, splintering his ankles and lower legs. Although his early recovery was good, his health began to decline after the accident. First there was minor gall bladder surgery, and later his doctor found he had an advanced case of colon cancer.

George Shoemaker was a very socially oriented person. Almost anyone could ask for a loan or a favor from him, says Carolyn, and get it. He made friends easily and kept them. As he lay dying, a man the family had never seen before knocked at the door. He wanted to visit George and remind his family of an event that had occurred long ago. The two men had been working on a scaffolding. The other man, who was on the next level up from George, slipped and fell. With his amazing strength, George reached out and grabbed the man, set him on the beam he was standing on, and said, "Going somewhere?"

George always kept an eye on Carolyn, whom he adored. One warm afternoon in 1953, Carolyn, Gene, George, and Toby, their Labrador retriever, were hoeing weeds around their strawberry patch. Unaware that Toby was right behind her, Carolyn swung the hoe and clipped the dog on his forehead. Although the dog was not seriously injured, everyone was appalled. Gene immediately berated Carolyn, but his father would hear none of it. "She didn't do it on purpose, Gene," George said. "Don't be that way!"

As George grew weaker in 1960, Gene and his family took advantage of every opportunity to visit Los Angeles. George had always enjoyed spending time with his grandchildren, talking and playing with them, but as he grew weaker he feared that they would remember him as sick and distant rather than as a helpful older friend. "They're not old enough to remember me when I was healthy," he lamented. Although George was never told of the unsuccessful results of his surgery, or of the poor prognosis, he undoubtedly was aware of his fate; however, he went along with the family's upbeat conversations and discussions about future plans. "It was difficult to speak of these plans," notes Carolyn, "when we knew they'd never come to pass."

Gene's father died in the late spring of 1960, at the young age of fifty-six. The family was stricken by this terrible loss—neither Gene nor Carolyn had ever experienced a death before. "The pain of losing someone," Carolyn noted, "was not something he could verbalize." Years later Gene wrote that "he was my father, and he was my hero. As a youngster I idolized him. In my youthful eyes, he could do anything. The passage of years has not changed that perception very much." [5]

George was delighted that by the end of his life, his only son had received his doctorate from Princeton. Gene thought it might have been the shortest thesis ever submitted to Princeton's geology department. Completing his thesis cemented his opinion that a thesis should not be the most important thing a scientist does, and that a ten-year research project that produces a tome that few will ever read should not be a science student's ultimate goal. Research for a thesis should allow for partial publication en route, and the thesis itself should be but a step in a grand process of learning. It is a sentiment with which, no doubt, his father would have agreed.

CHAPTER 8

A Dream Ends, A Dream Begins:
1960–1963

Goe, and catche a falling starre

—JOHN DONNE, *1633*

HALF A CENTURY after Daniel Barringer first started digging there, Meteor Crater was finally coaxed into revealing her secrets. The fieldwork on the Hopi Buttes Gene enjoyed so much was being replaced first with fieldwork around Meteor Crater, and later with a different kind of field study altogether—the close inspection of old photographs of the Moon. The days and weeks in the field with Carolyn and their young children were coming to an end.

EDWARD CHAO AND THE DISCOVERY OF COESITE

In the spring of 1960 an article in the *Journal of Geophysical Research* by geologist Joseph Boyd gave Gene's dream more strength. The *JGR* research built on a paper published seven years earlier by Loring Coes, who had used a large hydraulic press to squeeze a block of quartz and form a new, higher-density mineral.[1] Boyd established the minimum conditions of extremely high temperature and pressure under which this mineral, named coesite, might crystallize. The impact of a comet or an asteroid, Shoemaker thought, would offer an environment for such a mineral to form in the real world. Could coesite be the *sine qua non* of impact geology? Besides Gene, Edward Chao of the Smithsonian Institution was also working on the problem. Fetching a Meteor Crater sample on display at the Smithsonian Institution, he used an X-ray diffracto-

10. Ed Chao and Gene Shoemaker announcing the discovery of coesite at Meteor Crater, 1960. This seminal discovery showed beyond doubt that Meteor Crater was the result of an impact.

meter to detect the presence of coesite. On June 20, 1960, Chao announced the seminal discovery that coesite proved that Barringer crater was of impact origin. He then visited Gene and his colleague Beth Madsen—and demonstrated how to find the coesite mineral using X rays. Gene used the extensive collection of shocked rocks he had gathered from Meteor Crater, and detected coesite in some of them.[2] Although the discovery of coesite in the Canyon Diablo meteorite at Meteor Crater must be awarded to Chao, it was Gene who had pushed the research in that direction.

In July 1960 the discovery article appeared in *Science*, with Chao, Shoemaker, and Madsen all sharing credit.[3] The Shoemaker family was not home to enjoy it, however. Mourning his father's death, Gene suggested that his mother join them on a European driving trip to West Germany and hence to the Geological Congress in Copenhagen. On the way, he wanted to explore Bavaria's Rieskessel, a 24-kilometer-diameter basin with the town of Nördlingen at its center.

The Shoemakers drove through a summer rain and arrived at the site at sunset. The day's rain had stopped. "By the light of the setting Sun," Gene recalled, "we looked at the shock-formed rocks that were thought to be volcanic. I took one look at these rocks with a hand lens: no question these were impact rocks!" With clearing sky and the onset of darkness, the trio camped in a quiet wooded area. On the verge of a great discovery, Gene slept somewhat fitfully, wondering what this area might have been like eons ago as a monster fell out of the sky and hit there. After breakfast next morning, they made their way to the small town of Nördlingen, where Gene used his command of German to locate the post office. Once there, they sent two samples to Edward Chao in Washington via airmail.

As the Nördlingen post office prepared to send the samples flying for the second time in 15 million years, Carolyn suggested they visit St. George's Cathedral, a mighty structure that dominated the center of the peaceful village. As they looked up at the tall spires, Gene took out his field lens and studied the stone of the cathedral. He looked up at his wife and his mother and shook his head. The building was made of rock that included suevite, a mineral that Gene believed had been shocked and partially melted by the Ries impact. "And at that moment, I was the only geologist in the world who knew it!" Back in Washington, Ed Chao's X-ray diffractometer quickly confirmed Shoemaker's diagnosis by finding coesite within the suevite. This was a major find: the object that had fallen there was no building-sized rock but something the size of a village. The discovery sparked a flurry of activity among German geologists who had a field day with their newly

11. Aaron Waters and Gene at Meteor Crater.

recognized impact crater, and it set off a worldwide scramble to find other impact sites around the world. In Canada, an aerial search for impact craters had already led to the discovery of some promising sites, including a three-kilometer-wide crater a short distance from Kingston, Ontario, that was formed when something slammed into the Earth some 550 million years ago—one of the oldest events on record.

Years later Shoemaker revisited Nördlingen, this time to receive, along with Chao and a German geologist, the first Ries cultural prize. "It was so good to see that geology is part of their culture," Shoemaker noted.

Back from his highly successful 1960 journey overseas, Gene proceeded enthusiastically with his map of the Moon. These were very exciting days. Gene could taste his dream of heading for the Moon as he set to work on old photographs of the Moon. With the disbanding of Project MICE in 1958, Gene and Carolyn joined the Survey's Menlo Park center on California's San Francisco peninsula. Here Gene worked with a group of people interested in exploring the Moon and planets, and after a colloquium in early 1959, he began drawing a rough lunar geologic map to demonstrate that one could solve the stratigraphy of rocks exposed on the lunar surface.

Gene proceeded with his lunar fieldwork in much the same way he had out on the Colorado plateau. He couldn't do the fieldwork himself—not yet, at least—with hands, feet, geologist's hammer, compass, and pacer, but he could vicariously with his eyes. Some forty years earlier, on exceptionally steady nights when the 100-inch Hooker telescope at California's Mt. Wilson Observatory was new, Francis G. Pease had taken such high quality photographs of the Moon that they picked up craters as small as one kilometer in diameter. Shoemaker had enlargements made of the region around the crater Copernicus, a hundred-kilometer-wide feature that he thought resulted from an ancient comet impact. From these photographs his team made the first geologic map of a lunar feature. Besides Copernicus, the map showed a whole set of geological features in a region about the size of Arizona. Later he expanded the mapping into adjacent areas. Once completed, the project included fifty quadrangles covering much of the entire side of the Moon visible from Earth.

At the Eighth Lunar and Planetary Exploration Colloquium, held in Downey, California, on March 17, 1960, Gene presented results of his work on Copernicus. He demonstrated how an ejecta blanket (a covering of material thrown off during the later stages of evacuation after an impact and lying on top of much older blankets and other geologic structures) can provide clues to the relative ages of the craters. These blankets consist of lunar material hurled up and landing again, forming secondary craters that could also help establish the sequence of events during an impact. He suggested that small satellite craters, near the main site, were the results of ejecta hurled out from the formation of Copernicus. These secondary-impact craters, or secondaries, could easily be explained if Copernicus was the result of an impact.[4]

MENLO PARK AND THE BIRTH

OF ASTROGEOLOGY

When Gene and Carolyn moved to Menlo Park, he was set on two goals. One was to start a new science by founding a Survey branch of astrogeology, and the other was to put that new science into the nation's space program. While just a few years earlier, studying geology on the Moon was looked on almost as a crackpot idea, in the face of the developing space race, it was now taken quite seriously. Younger geologists like Jack Schmitt and Don Wilhelms saw the beginnings of planetary geology as their future, and they enthusiastically joined forces with Gene.

As a first step, Gene was successful in getting permission from the USGS director to launch an astrogeologic studies program at the Survey's Menlo Park office. Gene was fully immersed in his new work. "Perhaps most exciting for him," Carolyn wrote about Gene at the end of 1960, "has been the formation of an Astro-Geologic Studies Group within the Survey. Involved with its administration, he is busier than ever with moon studies"—leading Carolyn to think that—" 'lunatic' is a most appropriate name in this case."[5] Soon five geologists were on board. Gene saw Caltech, with its ready crop of young geologists anxious to try new adventures, as fertile ground for hiring members for his staff.

THE SHOEMAKERS GO TO WASHINGTON

On May 5, 1961, Alan Shepard became the second human to venture into space. His *Redstone* rocket soared off its launchpad and the *Freedom 7* capsule flew in space for a quarter hour before gently splashing down in the Atlantic. On May 25 President Kennedy announced that the nation was setting for itself a goal of landing a man on the Moon before the end of the decade. It was clearly time for Gene to position himself at the heart of the American venture into space. His dream of landing on the Moon was now thirteen years old, and it surely looked as though it might come true sooner than he thought.

In July Gus Grissom flew American's second mission. Sadly, the capsule was lost after the flight ended, though the astronaut was saved. The program was ramping up quickly.[6] To proceed with his goals, it seemed necessary for Gene to move his family to Washington for a time. While there, he could develop the contacts within NASA to ensure that science was placed prominently in the space program. At the same time, Gene could advance his goal of setting up a branch of astrogeology within the Geological Survey.

On February 20, 1962, John Glenn flew a successful three-orbit flight. Scott Carpenter followed two months later, and that fall, as Walter Shirra prepared to lift off into space for a six-orbit swing around Earth, and as Soviet missiles were being placed near launching pads in Cuba, the Shoemaker family arrived in JFK's Washington to work with NASA. The space agency was considering setting up a separate field center for geology. Gene helped change their minds to permit the USGS to handle the geology program, and according to Steve Dwornik, then program scientist for the Surveyor program, NASA management was very happy that Gene Shoemaker was there. "NASA used Shoemaker as a senior knowledgeable person with no axe to grind," says Dwornik. "As in private industry, in Gene, NASA had someone they could trust to suggest what science could be done on their space missions."[7] Gene's superior at NASA was a geologist named Oran Nicks.

The Washington year began well; Gene and Carolyn found that it was easy to make friends; in Washington people came and went so quickly that newcomers fit in easily. Washington, the Shoemakers quickly learned, also offered a host of new bugs and viruses that the children had not been exposed to. Christy and Linda each had scarlet fever, strep throat, colds, and a measles-like illness called fifth disease. Patrick passed his case of the mumps to Christy, who then developed encephalitis, and in the spring of 1963 Linda and Carolyn got bad cases of mumps.

Gene went through this depressing period well, at least until the onset of winter, when he suffered a strange loss of appetite, certainly unusual for him. His strength began to ebb until Christmas, when he was not feeling strong enough to enjoy this normally happy family time. Linda remembers "wanting to play with Dad, but Dad was too tired."[8] While at first Gene suspected he had caught one of his children's illnesses, he soon knew that his symptoms were entirely different. His skin grew dark, and he had frequent attacks of hiccups that were very hard to stop. After returning home it was all he could do to walk the few steps from the front stairs to the living room couch; he had no strength to involve himself in his family's evening activities.

Meanwhile, a family tragedy magnified the feelings of dejection that Washington offered the Shoemakers. One morning Patrick, then seven, left home to walk to school with his closest friend. Suddenly, Pat's friend inexplicably darted out into the path of an oncoming car. The driver couldn't stop in time, and Pat's friend was killed. This catstrophic event was devastating to an already unsettled family, and Gene's own illness and frequent absence limited his ability to be helpful at this time.

In the spring of 1963 the last *Mercury* flight coincided with a trip Gene took to Houston. It left him exhausted, but a visit to his physician still did not reveal the cause of his malady. With sadness and illness still gripping the family, Carolyn decided to take piano lessons to help relieve her stress. In June she and Gene decided to try to take a river trip, a total vacation for a few days. By the time they got off the river, Gene was so tired he could hardly move; his skin became even darker, even on the palms of his hands. Again, his doctor did not have any answers. By the fourth of July, the family decided to shoot fireworks, which were legal in Washington at the time. His weight was down to 128, and he looked, Carolyn remembers, like a refugee from a concentration camp. It was time to try another doctor. "Oh!" the new physician said, "you look like you have Addison's disease. I've seen two other cases like this, and of course they both died."

Named for Thomas Addison, the English physician who recognized it in 1855, Addison's disease is the result of the atrophy of the cortex, or outer layer, of the adrenal gland. Gene shared this disease with another Washingtonian at the time: President Kennedy had it as well, although the fact was not widely known. Before Gene and Carolyn got too discouraged, however, the doctor said that cortisone was showing very promising results for Addison's patients. When he entered the hospital for tests, he and Carolyn were saddened; the encyclopedias they had consulted said that this disease was usually fatal.

The day after he was hospitalized, he underwent several tests that confirmed the diagnosis of adrenal cortex shutdown. The following day he began taking cortisone tablets, and the day after that, says Carolyn, he was a new man. His condition improved dramatically by the hour, and he went home feeling stronger and more optimistic than he had felt in months. The next weekend the family went to the beach, Gene with more energy than anyone else! Addison's disease is a lifetime condition with no cure, but cortisone is an effective treatment. Gene would take these pills for the next thirty-four years.

By the end of July the family was finally over their season of misfortunes, and after nine months of illness, the final two weeks of health and energy seemed miraculous. But their time in Washington was over. Especially with the city's oppressive heat and humidity, the family knew it was time to move on. Despite his poor health, Gene had successfully launched a full Geological Survey branch of astrogeology and had persuaded the Survey to locate it in the northern Arizona town of Flagstaff. As the family pulled out of Washington, they cheered; it was time to head west.

In the spring of 1962 a holiday spoof made caustic fun of a scientist called Dream Moonshaker, haplessly standing on fresh lava pouring out of a fresh crater on the Moon's surface and trying to send a radio message that the Moon's craters were of impact origin. Gene didn't attend that presentation, but years later astronauts he trained stood on the Sea of Tranquility, which had indeed been hot lava 3.9 billion years earlier.

On the day of that spoof, Gene was visiting Flagstaff with colleague Dan Milton. It was an important trip; Milton made the suggestion that the branch of astrogeology belonged in Flagstaff rather than at Menlo Park, and Gene, who loved the small town on the Colorado plateau had the same thought, ran with the idea. It was the beginning of a plan to bring the new branch of astrogeology to the Colorado plateau city of Flagstaff.

Much as Gene loved the idea of living in Flagstaff and on the Colorado plateau, he also knew there were sound scientific reasons for setting up the Survey's astrogeology branch in the small northern Arizona town. Primary among those was the presence of that obviously astrogeological landmark, Meteor Crater, but the area was rich in volcanic features as well, particularly Sunset Crater.

Further, both Lowell Observatory and the U.S. Naval Observatory had facilities there. Shortly after the Shoemakers arrived in Flagstaff they discovered that scheduled airlines would be inadequate to handle the travel that the survey work would require. Anticipating that the Survey might buy its own plane, Gene and Carolyn planned to take flying lessons; Gene wanted to show that it was easier for a scientist to learn to fly than for a pilot to learn science. To accomplish this, the couple filled out routine application forms for pilot's licenses. Gene was turned down for a medical reason; some Addison's patients faint unexpectedly, and even though Gene's case was well under control and he never passed out, he was forbidden from piloting a plane. He was told that he could appeal this medical decision, but it did drive home the stark reality of his situation: if he couldn't get a licence to fly, how would he ever be allowed to go to the Moon? It wasn't until Gene was told he couldn't pilot a plane that the real import of his illness finally hit him. In 1948 his dream to go to the Moon began; in 1965 he had to accept the fact that his dream would forever stay a dream. "No—I'm not going to go to the Moon," he reminisced years later. "Just at the moment when I was at the head of the line to go to the Moon, my adrenal cortex shut down. I still dream of going there. But no—I just had to do something else."[9]

CHAPTER 9

Just Passing By on My Way to the Moon: 1964–1965

It is the very error of the moon,

She comes more nearer earth than she was wont,

And makes men mad.

—SHAKESPEARE, Othello, *1604*

EVEN AFTER he realized that the trip to the Moon would not be his, Gene Shoemaker never lost sight of his dream. He had been a part of Project Ranger's television team since 1960 and was intensely interested in Surveyor and Lunar Orbiter, the two other unmanned lunar exploration missions. Joining Gene on the Ranger team were Gerard Kuiper and Harold Urey, then two of the most knowledgeable people about the Moon. Back in 1961 Rennilson planned a trip north to Menlo Park to meet with Gene, and then east to Tucson, where Gerard Kuiper, science team leader for the Ranger project, was building the Lunar and Planetary Lab at the University of Arizona.

Rennilson's trip did not work out quite as expected. Gene, it turned out, would not be at Menlo Park at the time of their meeting, but instead would be off visiting Meteor Crater. Rennilson was delighted at this opportunity. He flew to Flagstaff, rented a car, and drove the hour and a quarter to the crater. From the bottom of the crater Gene saw JPL's point man, and he climbed out quickly.

What happened next was a unique event, something that did not seem earth shattering at the time but turned out to be in retrospect. Sitting together at the rim of an Earth-bound impact crater, the two

12. Gene in the field, 1964, age thirty-six.

men planned a scientific assault on the Moon. Looking across the expanse of the crater, Gene visualized how the United States would eventually land on the Moon. Rennilson had looked forward to this meeting—"Gene was thought of as brilliant," he said, "a wonder"—and he was not disappointed with his first acquaintance with the young geologist. Rennilson got the distinct impression that Gene, in those heady, pre-illness days of 1961, hoped to be the man to land on the moon. After that rimside meeting, Rennilson flew on to Tucson where he met Kuiper and his young associate Ewen Whitaker. These two were planning Ranger, a project that already was close to its first launch.

In 1963 Gene was offered a highly coveted position on the new Project Surveyor, as principal investigator of the television experiment, a task that would solidify his prominent role in the nation's space program. Where *Ranger* was intended to slam into the Moon, *Surveyor* would caress it, landing softly to begin a week long exploration. To begin the scientific preparations for *Surveyor,* Gene planned to meet with Jet Propulsion Lab's (JPL) Justin Rennilson, the liaison between the engineering team at the lab and the scientists, who were spread out across the country.

Rennilson recalls that Gene's thinking about the consequences of major impacts extended at least as far back as the early 1960s. Gene often talked of the Ries Basin as the result of a kilometer-sized comet or asteroid, large enough to create major consequences for the Earth's entire biosphere. What would happen, Gene asked, if a comet or an asteroid many times that size collided with Earth. Surely this happened many times in Earth's history. Could such an event 65 million years ago have caused the extinction of the dinosaurs?

Gene pushed his Moon dreams at every possible opportunity. One of those opportunities came at the University of California at Berkeley, where he was to give a colloquium that if successful might result in his being offered a professorship there. Gene worked hard on that colloquium, for he knew that he was being considered for a faculty position. While he enjoyed the USGS, Gene even then did not see it as his only world, and he at times wondered if an academic life, with its combination of research and teaching, would be preferable. Gene was wound up for his colloquium and talked avidly about what exciting geology might await the first visitors to the Moon. The session was a big success. The students were very enthusiastic and their questions went on and on. He returned to Menlo Park looking forward to a possible career shift. But soon he received a letter from Berkeley's geology department: Although his colloquium was a success, the faculty decided not to offer a position. His enthusiasm about going to the Moon was contagious, the letter opined, and the department feared that he might persuade Berkeley's impressionable young students that they may want to go there, too!

GETTING TO THE MOON WITH RANGER

Conceived by a group of engineers at Caltech's Jet Propulsion Laboratory, *Ranger* was ambitious almost beyond belief. "JPL was a really vigorous place," said Gene, recalling those heady days. "These people were going to build a spacecraft fast and send it to the Moon." Shaped like a hexagon, the early *Rangers* bristled with instruments for several experiments. There was a television camera, and a gamma-ray spectrometer to measure detectable radioactive elements on the Moon. Tucked inside was a balsa wood instrument container atop a retrorocket. The idea was that the balsa wood package would detach from the main spacecraft as it approached the Moon. The retrorocket would fire as long as it could, slowing down the balsa wood container and landing it on the surface. Then a small seismometer would right itself while an antenna pierced the encasement and started sending data to Earth.

The project seemed viable on paper, and *Rangers 1* and *2* performed successfully in trial runs. But 1962 would be the year of the failed *Rangers*. In January Gene left his home at Menlo Park and headed south for Pasadena. On January 26 the *Atlas* rocket roared from its launchpad, sending *Ranger* on its voyage. But a few days later the spacecraft sailed past the Moon, missing its target by some twenty-five thousand miles. On April 13, 1962, the engineering and science teams were ready to try again, and once again Gene traveled south to JPL. This time the *Atlas-Agena* rocket worked flawlessly, sending the craft on a direct course to the Moon. But this time the spacecraft failed to turn on its instruments before landing, and the $20 million craft, its balsa wood container still aboard, vaporized on impact.

Since *Ranger* 4 actually collided with the Moon, it provided an ideal opportunity to test an idea that Rennilson shared with Gene that Ranger would leave a visible flash as it impacted into the Moon. They calculated the mass of the spacecraft and used several telescopes, including the giant McMath Solar Telescope at Kitt Peak National Observatory, to see if the flash would be seen. The professional effort was accompanied by a large group of amateur astronomers who used their own telescopes to watch the Moon for signs of an impact flash. For many years prior to the *Ranger* flights, the Association of Lunar and Planetary Observers had sponsored a Lunar Meteor Program where its members would observe the unlit portion of the Moon on three days per month when the Moon was a thin crescent in the evening sky. Although several flashes were reported during the program, none was confirmed by a second observer.[1] Attempts to spot *Ranger 4*'s collision were also unsuccessful, since the craft sailed to the far side before impacting. Had it landed on the near side, and on the Moon's unlit portion, the flash against the darkness might have been detectable.

On October 8 a third attempt was started with the launch of *Ranger 5*. Of the original *Rangers*, this one came the closest to success. Though its instruments functioned, the craft missed the moon by just a few hundred miles—a very disappointing failure. Like *Ranger 3*, *Ranger 5* is now forever in an orbit around the Sun.

After three botched attempts to land a craft on the Moon with *Rangers 3, 4,* and *5,* the mood quickly soured. "Either the spacecraft was great," Gene lamented, "or the *Atlas* booster was great. Trouble is, we couldn't get a good rocket and a good spacecraft at the same time."

The *Ranger 5* failure was particularly disappointing because it had come so close to success. NASA and JPL decided to put the whole program on hold, rebuild the moon craft from top to bottom, and simplify. There was no question that the project must eventually succeed, for more than a year had gone by since President Kennedy had committed the United States to landing a man on the Moon by the end of 1969. Nevertheless, it was still the case that no one had any idea what the surface of the Moon looked like.

Engineering firms all over the nation were gearing up for mighty Project Apollo, and one of the most important considerations in the fall of 1962 was what kind of surface their craft would find. Getting high-quality pictures of potential landing sites became the top priority. So of all the changes planned for the new *Ranger,* the most important was in the camera design for *Ranger 6,* RCA supplied a system of six cameras set up so that at least one would record and transmit a partial frame right up to the instant of impact. This way, the last of *Ranger*'s images, the closest to the Moon's surface, would resolve regions as small as a dinner table.

The first casualty of this new thinking was the balsa wood instrument container, which was dropped from the *Ranger* redesign. That really was a shame, for this strange wooden ball had a rocket motor to slow it down for a reasonably soft landing on the lunar surface. The ball would land intact, it was hoped, and after bouncing a few times on the lunar surface, it would then activate its instruments and record all kinds of useful data. Of all *Ranger*'s aspects, the balsa wood container probably achieved the most press coverage. Designed for a period of lunar exploration, the *Ranger* project was now sorely needed as a preparatory mission for a future manned landing. There was really no single problem with the early *Rangers*; in fact, a closely similar version called *Mariner 2*, minus the balsa wood package, flew past Venus in December 1962.

The new and improved *Ranger* consisted of a simple cone-shaped tower with six television cameras. The science teams were disappointed with most of the changes, with the notable exception of the imaging team of Kuiper, Urey, and Shoemaker. Since pictures were so important, the imaging team now enjoyed top priority. In any event, four new spacecraft, *Rangers* 6 through 9, were assembled at JPL. At the end of January 1964, after a hiatus of more than a year, all was ready for the launch of *Ranger 6*. The excitement and confidence surrounding the early *Ranger* missions was replaced by a "cutting tension," as Gene noted at JPL. "The *Atlas* performed like a jewel," Gene remembers. "The *Agena* performed like a jewel. The spacecraft was fine. It was sending a clear signal as it made its final approach to the Moon. Everyone in the lab was huddling around the speakers, listening to the reports from the Goldstone tracking station. When the spacecraft was about a thousand kilometers from the lunar surface, the final critical signal was sent to turn on the cameras."

"No video."

For a second time, the signal was sent.

"Still no video."

For the next few minutes the room was silent except for the announcer repeating "Still no video." Then the spacecraft crashed into the Moon, its faint signal silenced. The control room at JPL, recalled Gene, was as silent as a crypt.

Ranger's four failures were extremely frustrating. Just after the failure of *Ranger 6*, there was a party for the mission's teams. The atmosphere was quiet and sad, except for two people, Zdenek Kopal and Gene Shoemaker. "Why be glum?" Gene complained. "*Ranger 7* is going to work!" A simple short circuit, it was concluded, cut off the cameras during the stress of launch.

Six months went by before *Ranger* tried again, this time in the form of a craft called *Ranger 7*. Once again, Gene joined an anxious team at JPL. Once again, the *Atlas* performed flawlessly, and the *Agena* successfully put the *Ranger* craft into a trajectory to the Moon. For the next three days the group waited as the spacecraft approached to within six hundred miles of the Moon; at that distance the cameras would turn on. Everyone in the room knew that the future of the Moon program rested on what would happen in the next few seconds.

Cautiously, nervously, an engineer dispatched the signal to turn on the cameras.

"We have video!"

It was an incredible moment. "Everyone was jumping up and down," Gene raved. "We didn't even know what the pictures were yet, but that didn't matter. There were pictures!" *Ranger 7*'s success was sorely needed after two and a half years of frustrating misses. On a very small scale, little *Ranger* was repeating a kind of episode that the Moon was very familiar with, an impact.

Ranger impacted near a ray from the crater Tycho, one of the Moon's most recent, large impact craters. Formed some 100 million years ago, the crater is the remains of an impact event of some long-gone comet or asteroid. As the incoming object struck, lunar rock shot out in all directions, landing again in a series of ray structures that stretch halfway around the Moon. *Ranger 7* was fortunate enough to come down near one of these structures, thereby providing our world with its first-ever view of the structure of a lunar ray. The ray consists of innumerable secondary craterlets that range in size, Gene saw, from doors to rooms.

By the end of 1964, learning about the nature of the Moon's surface had become *Ranger*'s main goal. When *Ranger 7* finally returned pictures, Kuiper suggested that walking on the surface of the Moon would be as easy as hoofing through crunchy snow. No more fears of spacecraft sinking in yards of dust. Urey said that the surface was gardened, or turned over and over by small impacts occurring over the millennia.[2] But before Gene returned to Menlo Park he was already trying to persuade mission managers to aim *Ranger 8* closer to the Moon's terminator, the moving zone of lunar sunrise or sunset, where lower sun angles would result in far better pictures. Some engineers argued that the contrasty light at the terminator would be insufficient for the spacecraft's cameras. But once *Ranger 8* was also a great success, Gene and his group finally prevailed for *Ranger 9*. Pointed directly at one of the most interesting regions of the Moon, the craft crash-landed inside the crater Alphonsus.

The brightest spot on the Moon, Alphonsus has always attracted both amateur and professional observers. In 1958 the Soviet astronomer Nikolai Kozyrev recorded a series of whitish glows in this crater, and his observations through a spectroscope showed evidence that gases were being emitted. *Ranger 9*'s pictures could answer the question of whether there was volcanic activity there. Arriving just after sunrise on the crater, *Ranger* snapped almost six *thousand* pictures before it vaporized in the crater upon impact. Although the pictures showed some evidence of ancient volcanic activity, the crater was quiet when *Ranger 9* crashed there, as it had probably been for millions of years.

GROUP FOUR GETS CHOSEN

Meanwhile, planning was intensifying for the crucial, manned phase of the Moon project. When Gene's friend and thesis advisor from Princeton was asked to chair the committee that would review the scientific qualifications of scientists who wished to join the *Apollo* astronaut corps, Gene had the feeling that his old dream was about to come true. Even if, as now seemed likely, this dream would not include him personally, it could still include him vicariously. "What I thought about in 1948 seemed to be coming to pass," Gene said. "Harry Hess chose to head up the committee himself; he asked several people, including me, to work with him. But when the time came for the committee to do its work, he was on sabbatical in England so he said, 'Here, Shoemaker, you chair the committee!' Instead of being there at the head of the line with my application to go to the Moon, I ended up chairing the committee that reviewed the other applicants." Out of about fourteen hundred applications for "group four," only some four hundred scientist-astronauts passed the initial physical exam provided by the Federal Aviation Administration. These four hundred were asked for further information, which included transcripts and recommendations, information that reduced the number to eighty. Next, the committee asked to see the research publications and a short essay about what the applicant would do if he had the chance to go to the Moon. Sixteen finalists went to Brooks Air Force Base in March 1965 to take an intense week-long physical conducted by NASA and the Air Force. Harrison "Jack" Schmitt was one of those. "Even though I made the final sixteen," Jack said, "I was pessimistic. I thought NASA would find something wrong with me, and because we were ten years older than the previous groups, NASA could reject the lot of us."

In May of 1965, just weeks after the first two-man *Gemini* spacecraft was successfully flown, Schmitt received a call from a consultant to NASA, hinting strongly that NASA believed he would have no problems in space. Three days after that, Alan Shepard, the first American in space, called Schmitt to ask if he would be interested in becoming a scientist-astronaut. With *Ranger* finally a success, and scientist-astronauts chosen, the Moon seemed finally in view for human exploration.

CHAPTER 10

Surveyor's Golden Years: 1966–1968

Four happy days bring in

Another moon; but O, methinks, how slow

This old Moon wanes! . . .

    Four days will quickly steep themselves in night;

Four nights will quickly dream away the time;

And then the moon, like to a silver bow

New bent in heaven, shall behold the night

Of our solemnities.

—Shakespeare, A Midsummer Night's Dream,

circa 1600.

O F ALL GENE's rich experiences, those occurring during the period between May 30, 1966, and February 21, 1968, were probably his happiest. He was principal investigator of the television experiment for Project Surveyor, a challenge that kept him at the forefront of lunar science and in the public eye during this critical time of preparation for the first manned landing on the Moon. *Surveyor* was well named; the project was a serious attempt to go to the Moon and kick the dirt in preparation for *Apollo*. The last three *Rangers* succeeded in their closeup investigations of the Moon's surface, but Project Apollo's appetite for new information about the Moon required new probes both for soft-landing and exploring missions, and for a detailed, close-ranged orbital survey.

Instead of bombarding the Moon, a new spacecraft was being built that would land not with a crash but softly. It was time to make the transition to a craft that would do everything the earlier *Rangers* tried to do and more, and would do it more elegantly. Instead of having a wooden box bounce around on the Moon, *Surveyor*'s Christmas-tree-like frame would land so delicately that it would barely disturb the soil it lighted on. Once it settled, it would then photograph its new home for more than a week before shutting down to rest for the long and frigid lunar night during which the temperatures would plummet to more than two hundred degrees below zero.

The original *Surveyor* was an ambitious craft laden with several cameras and other instruments, but because *Rangers 3, 4, 5,* and 6 failed in their attempts to study the Moon, the Surveyor project was scaled down to its barest bones. "They threw off all the instruments," Shoemaker lamented, "leaving only a single camera atop the 'tree' as an engineering test." This left Gene wondering how he would ever get his first closeup view of the lunar surface. He spent considerable effort in experimenting and planning how *Surveyor*'s single camera would do its work once it finally landed.

In February 1966, the Western world got an unexpected lesson in what the Moon looked like. The Soviet spacecraft *Luna 9* set itself down on the Moon's surface much as the original *Rangers* had tried to do. Although the Soviets hadn't intended it that way, Britain and the United States spied on this craft thanks to Sir Bernard Lovell, the famous radio astronomer, and the giant dish of the world's largest radio telescope he directed at Jodrell Bank in England.

Jodrell Bank has an interesting and sometimes covert history with Soviet lunar probes that began with *Lunik 2* in 1959. The Soviets, it appeared, wanted the Western world to track the probe to prove that it actually had gone to the Moon, and Jodrell Bank's antenna was the most powerful on Earth. "I found on the telex machine," Lovell said about that time, "the long message from Moscow giving complete details [about its mission]—not only the transmission frequencies of this *Lunik*, but also its positions calculated for the latitude and longitude of Jodrell Bank. This, of course, was a clear indication that we were meant to do something about this."[1]

Seven years later, the Jodrell Bank astronomers again pointed their huge dish-shaped antenna to the Moon, this time to examine the signal from *Luna 9*. "When *Luna 9* hit the Moon," says Lovell, "the signals ceased abruptly, and we thought that was the end of the affair. But to our astonishment the signals reappeared, and this was facsimile transmission which in those days was used for transmitting photographs by newspapers. We had no facsimile machine, but fortunately it was the *Daily Express* who responded and immediately sent down the necessary equipment. . . . We saw before our eyes the astonishing thing—actual rocks of the Moon appearing. These were the first pictures transmitted from the Moon."[2] The Jodrell Bank team secured a good, though somewhat distorted, picture that showed a surface strewn with rocky rubble. Meantime the *Daily Express* took full advantage of its coup: "From *Luna 9* to Manchester," its headline blared, "the Express Catches the Moon."[3]

By the time *Luna 9* stopped transmitting on February 6, 1966, Gene had in his hands a firsthand view of what he called the lunar regolith. At the time, most geologists casually defined *regolith* as "the layer or mantle of loose, incoherent rock material, of whatever origin, that nearly everywhere forms the surface of the land and rests on the hard or 'bed' rocks. It comprises rock waste of all sorts, volcanic ash, glacial drift, alluvium, wind-blown deposits, vegetal accumulations, and soils."[4] For Gene, this first closeup revealed the nature of the Moon's layer of surface debris or soil. On the Moon, the regolith had a much simpler origin than on Earth, since the only force working on it was the slow, careful punching out of the rock by tiny meteoroid impacts.

Three and a half months later, Gene left for JPL and the mission of *Surveyor 1*. He told Carolyn to expect him back soon, because this first mission was testing so much that was new. This mission was so much more complex than the last *Ranger* probes that it seemed prudent to consider it a simple shakedown; the engineering and science teams would learn from it, and hopefully might succeed in soft-landing a craft on the Moon by *Surveyor 2* or *3*.

Surveyor 1: A Shot in the Arm, and to the Moon

"Of all the people on the Surveyor project," Gene said with not too much exaggeration, "only three of the engineers were confident that the craft would actually make it to the Moon. They had really worked that spacecraft over."[5] So it was that on May 30, 1966, *Surveyor* soared aloft aboard an *Atlas-Centaur* rocket. The launch was the first step in a long process during which a single problem could doom the journey. The *Atlas* performed exactly as planned. Next, the *Centaur* fired on schedule. This second stage was an experiment in itself, a rocket designed to fire once to ease its payload into Earth orbit and then to fire a second time to increase the spacecraft's speed to hurl the craft to 25,000 miles per hour, the velocity needed to break the bonds of Earth and head off to the Moon. It was the *Centaur*'s first operational flight, and it performed beautifully.

Now cautiously optimistic, Gene and his team watched the flight phase from Earth orbit to the vicinity of the Moon go well, and they assembled at the control room at JPL with the craft still operating. As the craft approached a critical distance of sixty miles from the Moon's Oceanus Procellarum, it was still racing along at 6,000 miles per hour. The retrofiring engine turned on exactly as scheduled, slowing the craft to 250 miles per hour; by this time tension in the control room could be cut with a knife: if anything could go wrong, especially considering the failure scenarios with the Rangers, it was likely to happen within the next few minutes.

The spacecraft's main engine fired for forty seconds, slowing the craft down. It shut off exactly on schedule. Surveyor was now falling moonward at 250 miles per hour, using an onboard altimeter to check its progress. After the retrorockets fell away, three smaller rockets took over, firing until the craft slowed almost to a stop just thirteen feet above the surface of the Moon. To leave the surface below as pristine as possible, these rockets then shut down. *Surveyor* fell quietly to a landing near the crater Flamsteed, the strong shock absorbers in its landing legs absorbing much of the shock.

According to all the telemetry, the spacecraft was on the Moon, its systems operating perfectly. In Mission Control, the group was astounded.

"It really landed! We were still communicating with it!" Gene was elated. "It was the most surprised bunch of people you ever saw!" But before the images could start rolling, it was first necessary to check out the spacecraft. As Gene and his colleagues waited for what seemed like forever, the engineering team checked out all the moon probe's systems. Finally, the camera was turned on. When it took its first picture of its landing site, Gene knew he was in for a busy week. "I slept about two hours over the next five days," Gene recalled abouat the incredible week that he had watched the Sun climb over the craft's landing site.

The many science-team training sessions that Gene organized in Flagstaff finally paid off. As each image of a portion of *Surveyor*'s landing site was transmitted from the Moon, the team tried to place it correctly into a mosaic showing the entire site. The team was awake and mapping out the landing site for at least thirty-six hours after the landing. Even though Mission Control had some rooms upstairs for quick catnaps, there was little sleeping done during that thrilling week. At the end of the week, Gene realized that a preliminary report was due immediately. So instead of going to sleep, he paced back and forth, with a microphone, and dictated the draft of the entire report.[6]

It was a very satisfying time. Without ever leaving the trappings of the newly completed Spaceflight Operations Facility at JPL, Gene was figuratively standing that week on the very surface of the Moon. The tiny camera atop the Christmas tree of *Surveyor* showed a regolith that was a blanket of material that had been thrown off by ancient impacts. The surface at *Surveyor*'s camping site was a sea of impact craters ranging in size from some thirty meters across to just a few centimeters. Don Wilhelms, one of the geologists who lived those days in Mission Control, wrote: "People are dashing about while never-before-seen views of outer space are flashing onto the monitors. In the 1960s and 1970s the central room[s] . . . were full of highly competent engineers knowingly contemplating their computer screens. The back rooms were equally full of scientists speculating on the meaning of it all. The rooms were windowless; night and day were identical, and the excitement went on 24 hours a day as in a Nevada casino."[7]

Surveyor could not, however, continue its work indefinitely. The Sun was setting on the craft's landing site, as seen by the lengthening shadows in *Surveyor*'s later photography. It was time to shut *Surveyor* down to prepare for what was to come, a two-week-long night during which the temperature would drop to some two hundred degrees below zero. Back in warm Pasadena, Gene was more tired than elated. He realized that despite its load of hardware, *Surveyor 1* had just passed one critical test: the spacecraft had landed, as he fully expected, on a solid lunar surface, and it hadn't sunk helplessly into many feet of dust.

Even though it made little geologic sense, scientist Thomas Gold expounded his theory that the Moon's surface dust would sink a spacecraft. Before *Surveyor*'s successful landing of *Surveyor 1*, the "Gold dust" theory became a controversial issue, if not to the scientists, at least to the press, to whom Gold argued his theory forcefully. In the science of *Apollo*, however, the idea of "Gold dust" was a red herring that took time and resources away from real science needs. The issue did lead to a debate at Caltech moderated by the famous physicist Richard Feynman. Gold presented his arguments. Even though Gene was not there, Feynman made it unnecessary for anyone to argue the other side. "That was an interesting story," Feynman said once Gold was finished, "with good sound physical arguments. But I'm afraid I can't agree with you."[8]

"There was a difference," says Steve Dwornik, a colleague who attended the debate, "between some theoreticians who needed recognition by the public and Gene Shoemaker, who went out there with a geology pick in his hand and cracked open the rocks and found out what there was. He dirtied his hand and ripped his knuckles and carried a geology pick. He was not looking for self-acclaim."[9]

Surveyor's preliminary report, however, contained little reference to Gold dust. Pacing round the room as he dictated, Gene finally completed his report and returned home to his family. A few weeks later, he set out on a return trip to southern California for

Surveyor 2. But even though Gene set his own travel course correctly, the spacecraft did not. A victim of a faulty midcourse correction maneuver, the second *Surveyor* did not make it to the Moon intact. The mission would have been interesting if it had—its older brother, *Surveyor 1*, was still functioning, and since the command frequencies were the same, the earlier craft would have tried to respond to the same commands at the same time.

SURVEYOR 3 AND THE REGOLITH

In April 1967 *Surveyor 3*, the mission that scouted the site for *Apollo 12*'s landing two and a half years later, landed near the eastern edge of Oceanus Procellarum, the Ocean of Storms, and right in the middle of a 150-foot-wide crater. The good news was that *Surveyor*'s camping site was an ideal place to study the surface of a crater floor and surrounding walls. The bad news: the camera couldn't see beyond the crater walls! Surveyor's myopic view was, however, augmented with a new device, a "soil-mechanics surface sampler." This device, also called a "scratcher arm," dug four trenches, broke up rocks, and tested the strength of the landscape. Thanks to this addition and the craft's wealth of some six thousand pictures, *Surveyor 3* allowed Gene and his team to nail down their interpretation of the regolith around the site.

On April 24, 1967, *Surveyor 3* saw and recorded an event never before seen by humanity: A total eclipse of the Sun *by the Earth*. The camera took two sets of pictures of the event, but the exposures were too short to detect the Sun's faint outer atmosphere, or corona.[10] A solar eclipse on the Moon corresponded to a lunar eclipse on Earth that day, which also corresponded to the night of full Moon and the first night of Passover. On that day also, the Soviets tried to reenter the race for the Moon after a 25-month lapse in manned launchings. Troubles beset their craft, which was brought down in less than a day. The parachute opened, twisted, and tangled itself; the spacecraft crashed, killing its cosmonaut. On the day that *Surveyor 3* recorded Earth's first view of a solar eclipse on the Moon, the USSR essentially lost the race there.

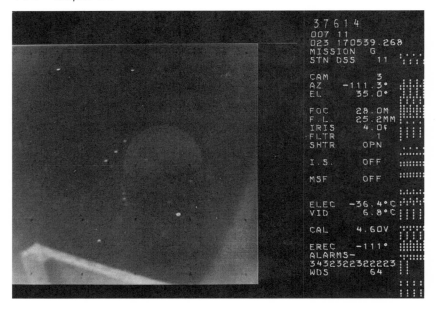

13. The first photograph of a constellation taken from the Moon.
Photo courtesy Justin Rennilson. NASA photo.

FAMILY LIFE DURING THE SURVEYOR YEARS

The fourth *Surveyor* snatched defeat from the jaws of victory by
mysteriously ceasing its transmissions just two and a half minutes
before it touched down in the Sinus Medii. Gene came home early
and disappointed, but his family, who had almost forgotten what
he looked like during this hectic period, was glad to see him and
somewhat relieved that they'd have an extra week with him. *Surveyor* and *Apollo* were Gene's life, and they kept him increasingly
occupied by 1967. It was a pattern that reflected stress and difficulty for everyone in the space program. The sixties were happy
years filled with hope and effort; the nation was preparing to send
a man to the Moon, and Gene was a big part of that effort. For
Carolyn and the children, however, it was a time of frustration.
"Gene would pack his bag and get on the train heading for JPL,"
Carolyn remembers. "There was so little news about these missions"—especially to a family without a television set. In Arizona,

William Graves Hoyt, a reporter for the *Arizona Sun*, kept the Shoemaker family up to date on what was happening more than Gene's occasional telephone calls did.

"Gene knew that I was awfully restless during those years," Carolyn says, remembering that period. "I tried not to show it, but when you live with someone the feeling comes across. We were very close even though we were apart so often. I believe he knew that I was not happy. I wanted to be with him but yet I couldn't. I saw marriages dissolve. The men were so committed, all of them, to the program. Then I asked myself, 'Do I want Gene as a husband part of the time, or would I want someone else all of the time?' 'No! I'll take Gene on any terms.' "[11]

The family's memories of Gene focus on the holidays, when, of course, the scientist was home. But home was also a chance to catch up on badly needed sleep. On Christmas morning, Carolyn had to restrain the three eager youngsters until Gene woke up at around eleven o'clock. At other times, Carolyn was active in the League of Women Voters, and in the language committee of her children's school. The family learned to ski—everyone, that is, except Gene, who believed that skiing was unnecessarily dangerous, and that a skiing injury would adversely affect his ability to do further fieldwork, an activity that in these space-mission days seemed a part of his past, but that he longed to renew.

SURVEYOR 5 AND 6: PREPARING FOR APOLLO

Surveyor 5 was sent aloft in September 1967. The craft flew perfectly until its rockets shut down fifteen feet above the wall of a crater not much bigger than itself. The craft fell to the crater wall, almost fell over, and slid three feet while gouging out trenches. Mission control had no notion of *Surveyor*'s adventure until it sent its first picture back. Perched at a twenty-degree angle on the slope, the spacecraft survived thanks to a design feature—which had worked during Earth tests and now was working in actual use on the Moon—that prevented it from tipping over if it set down at an angle that steep.

On November 7, 1967, *Surveyor 6* left for its lunar voyage. Gene's team was disappointed in the choice of site; by this time they had hoped to send the craft to the Fra Mauro highlands north of Mare Nubium, the Sea of Clouds. There the geologic bonanza was greater than the already explored plains. However, the craft explored another potential *Apollo* site, the small plain called Sinus Medii that *Surveyor 4* had missed—right in the center of the side of the Moon visible from Earth, a place Gene loved to call the Moon's belly button. *Surveyor 6* performed a fancy and sophisticated photographic trick. After its main photographing mission was done, the camera photographed some special targets. Then the craft's rockets fired just long enough to push it a few yards to a new location. As soon as the move was completed and the crafts systems checked out, the camera reshot the same scenes as it had earlier, this time from the new angle. Thus the first stereographic view of features from another world were taken. The stereo pictures provided a sense of depth that would be very useful when working out the distances and heights of the surrounding terrain.

By the end of 1967, Project Surveyor had sent four craft to successful soft landings and photographic missions to the Moon, by any definition a highly successful effort. NASA's Moon mapping project, called Lunar Orbiter, was also quite a success by that time, with two craft having flown in orbit around the Moon. Though Gene had had an intense early interest in the Lunar Orbiter program, he had found that he had had no time to concentrate on the project. Thus, colleague Jack McCauley eagerly accepted Gene's invitation to take over the USGS share of the project, which eventually flew a total of five spacecraft into lunar orbit without a single failure.[12]

Science and the Last Surveyor

With just one *Surveyor* to go, Gene and his colleagues wanted to try something different. They proposed an ambitious ride to the large crater Tycho, deep in the Moon's southern highlands. Lunar Orbiter had taken pictures that indicated the Tycho area to be very rough, but the potential payoff was tantalizing. Tycho is almost sixty miles wide, and its walls almost fourteen thousand feet high—an incredible structure built in a few minutes! At the time Tycho was formed, probably by the impact of a comet, dinosaurs were in the middle of their dominance here on Earth. (However, this chronology would not be known until after the return of lunar samples by *Apollo*.) Those creatures could have seen this comet vanish in a flash of blinding light as it crashed head-on into the Moon. The impact raised a cloud of dust, as huge chunks of material tore out of the surface and flew as far as halfway around the Moon before crashing again to form several series of secondary craters, which still can be seen as long "rays" stretching around the Moon. (These rays are best viewed through a small telescope near full Moon, and especially during a penumbral eclipse of the Moon.)

With all of *Surveyor*'s goals met at the end of 1967, even the people in Project Apollo were satisfied with its accomplishments. The final *Surveyor* was a spare, and the project managers decided that it could be used for science. To Gene's surprise, the *Surveyor* management welcomed the idea to close *Surveyor* with a dashing finale to Tycho.

Surveyor 7 launched on January 7, 1968. Before liftoff, Gene decided that he did not want to have his view of the launch constrained by what he thought was an arbitrary rule that stipulated a specific minimum distance for viewers. With Henry Holt and Ray Batson, Gene sped to a vantage point less than a mile from *Surveyor*'s *Atlas Centaur* launch vehicle on pad 36A. As they approached the massive launch complex, they saw the rocket that carried their hopes for a visit to Tycho. They found a beautiful site with a good view of the launch complex and quietly stood around waiting as the countdown progressed toward launch. As zero hour approached, however, security guards found them. They patiently listened to Gene's explanation of his role in the project and his desire to watch the launch, then informed the group that they were too close. The police escorted them out of the area; Gene drove quickly to a roadside spot, stopping just in time to watch the launch.[13]

Surveyor 7 landed quietly on a gentle five-degree slope on Tycho's northern rim. Its first picture revealed how precarious its landing was; the craft stood in the midst of a field of boulders, many of which could have punctured the craft or tipped it over. The site offered a great variety of rock types, including rocks crushed and deformed by the high pressure of impact. Gene was elated: "Tycho offered a much younger surface," he said, "than anything else we had seen." It also showed how the lunar regolith varies widely; *Surveyor 6*, which landed on an ancient lava bed, sat atop a regolith built from countless impacts over eons of time and was more than ten meters thick, while the Tycho regolith explored by *Surveyor 7* was less than fifteen centimeters thick. The thickness offered a way of estimating age; the thicker the regolith, the older the surface. Gene's interest in the regolith had a powerful payoff in terms of planning for *Apollo*. Older regoliths, which have been progressively built up and worn down over time, are smoother, and therefore safer for manned landings.[14]

By the time *Surveyor 7* had sent its last picture, on February 21, 1968, Gene, along with the rest of the nation, was ready for *Apollo*.

CHAPTER 11

One Giant Leap: 1968–1969

I believe that this nation should commit

itself before the decade is out, of landing a man on the

Moon and returning him safely to Earth.

—JOHN F. KENNEDY, *25 May 1961*

ALMOST SEVEN YEARS after Kennedy's challenge, his nation was finally ready to send humans to the Moon. At least it was, if the unmanned missions were any indication. *Lunar Orbiters 3, 4,* and *5* flew in February, May, and August of 1967—every mission in that program successful. Five of seven *Surveyors* were also successful. In just a few short years, astronomers and geologists had learned more about the Moon than they had in all of civilization, and so from a scientific point of view, we were ready to send a person to the Moon. But the manned program in early 1968 was a different story. A year had passed since Friday, January 27, 1967, when Gus Grissom, Edward White, and Roger Chaffee were killed when their spacecraft caught fire during a launchpad test. Although the *Surveyors* and *Lunar Orbiters* continued their work during the hiatus that followed, manned flights did not resume until well into the fall of 1968.

Considering what was happening in the United States in 1968, it seems amazing that there was any interest in a space program at all. At the end of January, the Tet offensive from North Vietnam intensified both the Vietnam War and the opposition to it. At the end of March President Johnson announced that he would not seek or accept the nomination of his party for another term as president.

A week later the assassination of Martin Luther King Jr. led to a week of rioting across the country. Two months after that, Robert Kennedy was murdered in Los Angeles.

MOON RIVER

For the Shoemaker family, the year began by helping Flagstaff pull out of a record ten feet of snow in seven days. Gene was in California when the snowfall began before Christmas, and when roofs began to collapse, he decided he'd better go home. Gene was able to catch a plane to Phoenix, where he rented a car and managed to make it to the town of Sedona and the beginning of the final drive uphill into the heavy snow. Highways and airports were closed; trains were not running, but the legendary Shoemaker luck held: he found a snowplow clearing the road, and followed it to Flagstaff. With snow still heavy on the ground, later that month the family moved into their new home—a partially built "ten-year project" with its lower floor almost completed and a magnificent northside view of the San Francisco Peaks.

With Gene's role on the Surveyor project nearing completion and the *Apollo* Moon landing still a year away, Gene decided to do something very different during the summer of 1968. He planned and led nothing less than a reprise of John Wesley Powell's pioneering expedition ninety-seven years earlier, down the Green and Colorado Rivers. Beginning in Green River, Wyoming, and stretching through the Grand Canyon, the trip began in July and lasted until September, and it succeeded in taking his mind away from the Moon. The expedition was intended as a scientific study of what changes had occurred in the previous century along the river system, and it evolved into a book written with Hal Stephens. Taking much longer than the expedition, the book finally appeared in 1987.[1] Even this project relates to Gene's overall view of the importance of the new catastrophism in geologic thought. In the Canyon, flash floods and rock slides have the ability to transform the landscape almost instantly in places, whereas in other places, the canyon landscape is virtually identical to its appearance a century earlier. The Earth's landscape changes by rare, local "catastro-

14. Gene, with son Pat, at the start of the Powell Centennial river trip. Hal Stephens photo.

phes," more than by the relentless, uniformitarian, gradual erosion that geologists had long envisioned.

That expedition was a good example of the way Gene loved to relax. Even though he appeared on television many times, for example, he and Carolyn rarely watched television. For Gene, the rigors of a long river trip constituted a perfect R and R for the busy scientist who seemed to bear the geological responsibility for the Moon effort on his shoulders. Years later the Shoemakers' field trips to Australia would also be precious times away from the office, mail, telephone, and other distractions.

It was probably just as well that Gene was on the Colorado River and out of touch with what was being played out at the Manned Spaceflight Center in Houston. There, a disturbing idea was taking form. Instead of permitting both astronauts to leave the Lunar Excursion Module, or LEM, and walk on the surface, one would always stay aboard. The other would leave, collect some random

15. Contemplating a new day on the Powell trip. Hal Stephens photo.

samples, and get back in, that is, no time for geology. In the absence of Gene, geologist Wilmot Hess led the effort to keep the geological program alive. The goal of the science cutback, according to Swann, was not to remove all science from the *Apollo 11* mission, but to remove geology as an experiment. But, since the ALSEP was flying aboard *Apollo 11* anyway, Gene was subsequently able to get the geology investigation reinstated.[2] Thanks to his efforts, a compromise was reached: there would be one extravehicular activity or EVA, not two, but both astronauts could take part and do a field traverse.[3]

OVER THE MOON

During the time that the fate of *Apollo 11*'s geology experiment hung in the balance, Chicago was totally disrupted when police and protesters surrounded the Democratic National Convention

16. Gene (on right) at the end of the Powell Centennial river trip. He is sporting the "special award" for swimming through Lava Falls. L-R: Hal Stephens, Maury Brock, George Ogura, Dave Gaskill, George Simmons, and Henry Toll.

there. And at about the same time, NASA learned that the Soviets were quietly planning to bring their own Moon program back to life with a manned flight around the Moon at the end of 1968. In October, *Apollo 7* made a completely successful test of the redesigned *Apollo* command module, and NASA boldly decided to go where no manned craft had gone before—to orbit the Moon and return to Earth.

17. Fabricating a "rolling frame" for riverboat trip to recover the camera stations at the Second Powell Expedition. Spring 1968.

In November, a week after Richard Nixon was elected president of the United States, the Space Agency officially announced that *Apollo 8* was heading to the Moon. In a tumultuous year, the decision rocked the country like a shot of adrenaline. On December 21, the second-ever flight test of the monstrous *Saturn 5*—a rocket as tall as a thirty-story building—roared away from Cape Kennedy, with Frank Borman, Jim Lovell, and Bill Anders aboard. When the *Saturn 4B* third stage finally shut down after sending its precious cargo toward the Moon, the lightly tested powered-flight phase was over, to everyone's vast relief. On Christmas Eve, as the three astronauts circled the Moon, Frank Borman read from Genesis, and talked about the "Good Earth" the astronauts could see through the spacecraft window, to an audience of millions watching on television as the words "Live from the Moon" flashed at the bottom of the screen.⁴ *Apollo 8* had a tremendous healing effect on the "Good Earth," and its astronauts were named *Time*'s "Men of the Year."⁵

WHEN THE MOON HITS YOUR EYE, THAT'S AMORÉ

As these events were taking place, Gene was continuing his own preparations for the Moon. In 1965, while *Surveyor* was completing its successful program, he had been appointed principal investigator of the field geology experiments for Project Apollo. It was a grand challenge and put him in direct involvement with the astronauts and at the center of one of the most prestigious national scientific efforts in two centuries. Although a few of the astronauts seemed bored with the geology, others were inquisitive and interested. Gene's goal was to get the astronauts to share his enthusiasm for the geology. If they were excited about it, he reasoned, they would absorb it better. Back in January 1963 Gene had taken the nine astronauts of the Gemini Project on the first geological field-training mission. The experience lasted just two days, but it included a visit to Meteor Crater and the nearby volcanic Sunset Crater, and nighttime observing of their ultimate target, the Moon, from the Lowell Observatory. He would carry this enthusiasm through his work with the astronauts for *Apollo 11* and *Apollo 12*, and in preparations for the ill-fated *Apollo 13*.

As training proceeded, Gene and his team at the USGS modeled a landing site on the Moon. Standing on the crater-strewn surface was a full-size mockup of the Lunar Excursion Module, arranged so that the astronauts could practice good field techniques. "Some of those test pilots were very good observers," Shoemaker remembers, adding that their flight training and alertness had given them the potential to be ideal field geologists—if they could get sufficient training. During this period Gene tried on the *Apollo* space suit and the rocket pack to see if the astronauts could do what he expected of them while wearing the pack. He never used the pack, or flew in the suit as an astronaut, but pictures of the suited-up

18. Gene tries on a spacesuit, with a mockup of the Jacob's Staff, just in case. He wore it to test astronaut reactions to the items he wanted them to carry on the Moon. Photograph courtesy NASA.

scientist did create that impression at times. Under the direction of Gordon Swann, the astronauts also trained in places as diverse as Meteor Crater, Grand Canyon, and the volcanic islands of Hawaii and Iceland.

In one of Gene's early astronaut training trips into Meteor Crater, he was warned that there were more rattlesnakes that year than usual. "Oh, no," Gene argued, "there are no rattlesnakes out here! Rattlesnakes are notably scarce throughout the Colorado plateau." They began descending the path now known as astronaut trail. Not far from the bottom of the crater, Jim Irwin, who would later take *Apollo 15* to the Moon, encountered a three-foot snake buzzing at him. "Hey Gene!" he yelled. "There's a snake down here, and he's making a funny noise at me, but I don't know what kind it is!" Shaking his head, Gene hurried down, stared at the rattler, and exclaimed, "I'll be damned!"

"Gene never saw rattlesnakes," explains Gordon Swann, "because he was always walking so fast. They're so well camouflaged that when they buzz they're already within striking distance. By the way," Swann added, "keeping up with him physically was a lead-pipe cinch compared with trying to keep up with him mentally."[6]

The early training sessions gave Gene some ideas as to what tools and implements the astronauts should bring with them for their lunar exploration. Even a simple geologist's hammer, the most common and essential geological field tool, was considered impossible. Gene realized that just hanging a hammer, along with a Brunton compass and pacer, to an astronaut's suit was out of the question once he tried on the *Apollo* moon suit and found it too bulky to handle any conventional geologic tool. To correct the problem, he and others came up with an ingenious device called "Jacob's staff" that the astronaut could carry like a cane. Housing a TV camera, leveling devices, antenna, and transponder, the device was very simple. All the astronaut would have to do was point it and pull a trigger; the transponder would be tracked by the LEM; the camera would take a picture. As mission managers rejected even this idea, Gene began to suspect that *Apollo* would become a mission of engineering rather than of exploration. He feared it would be a more complex version of the *Mercury* "man in a can" program

19. Gene points at an astronaut near a model LEM on a model Moon.

where the astronaut went along for the ride while ground controllers and simple computers gave all the instructions. As a field geologist, Gene believed that *Apollo* could be both for engineering and for exploration. Some engineers and managers agreed with him; the astronauts would get to use a gnomon, a device for establishing scale and orientation that consists of tongs attached to a rod.

On July 16, 1969, three of Gene's field geology students—Neil Armstrong, Buzz Aldrin, and Michael Collins—sat down in contoured chairs and waited to begin a field excursion of their own. In a roar of millions of gallons of burning kerosene, the *Saturn 5* rocket beneath them surged to life and bore the three men, and the hopes of our time, away from Earth. Gene and Carolyn were at Cape Kennedy watching, along with other notables like Charles Lindbergh.[7] As Armstrong, Aldrin, and Collins left Earth orbit and headed moonward, the flight was proceeding as though manned landings on the Moon took place every day. Hovering in lunar orbit, astronauts Neil Armstrong and Buzz Aldrin entered *Eagle*, their lunar module, and began their tricky descent to the lunar surface. "We had carefully mapped every crater in the landing strip on Mare Tranquilitatis," Gene noted. "It was an ellipse about ten kilometers long by two kilometers wide." But although *Lunar Orbiters 4* and *5* had recorded some gravity anomalies in the Moon—the mascons or mass concentrations associated with the Moon's huge lava-filled basins—*Apollo* planners had ignored them. As a result, instead of heading for the center of the ellipse, *Apollo 11* was about to land near the far end—right down in one of the largest craters in the whole ellipse.

"Every *Surveyor* that made it to the Moon—blind—landed safely," Gene said. "But the first time we had a man in the spacecraft we really needed him." When the crater was formed, it ejected tons of rocky debris—huge boulders that a billion years later would have caused the *Eagle* to tip over and crash. As Neil Armstrong urgently steered the craft away from the crater, Buzz Aldrin noted his progress, both downward and across:

Houston: 60 seconds (of fuel left)
Altitude 1600
1400
400 feet, 8 forward
300 feet, 47 forward
13 forward
11 forward
down one-half; picking up some dust . . .

4 forward, drifting to the right a little
(Houston, urgently): 30 seconds!
4 forward, 40 forward

With less than half a minute of fuel left, Armstrong set *Eagle* between two of the crater's rays.

OK, engines are off.

Gene Shoemaker, his science team, everyone in the Apollo program, and one-sixth of the world's population heard the next words from the LEM:

Tranquility Base here. The Eagle has landed.

Only a few hours later, Armstrong stepped down the ladder and onto the surface of the Moon. His words, "That's one small step for a man, one giant leap for mankind," marked the start of the most unusual geological field traverse ever undertaken. Five months before time ran out on President Kennedy's commitment, two American astronauts, trained in the methods of field geology, stood on the surface of Mare Tranquilitatis near Crater Sabine.

Armstrong and Aldrin remained on the surface at Tranquility Base for more than two hours. Armstrong noted big boulders more than two feet across, similar to the ones that had almost scuttled the landing attempt. He thought they were basaltic and added that "they have probably 2 percent white minerals in them, white crystals. And the thing I reported as vesicular before, I don't believe I believe that any more. I think that small craters—they look like little impact craters where B-B shot has hit the surface."[8] Armstrong's ninety-minute moon walk was very impressive—in Gene's view it was one of the best of the entire Apollo program. "He saw more stuff, and he made more pertinent observations, in the precious little time he had on the surface, than many of the astronauts who followed him." Armstrong's descriptions were lucid and accurate. Carolyn remembers her husband's feeling that Neil's performance was an enormous reward for time and effort.

For Gene and all of us, the return of the three explorers was a great relief, and we felt satisfied that a new chapter in human exploration had successfully begun. It was hard to be unemotional about the plaque the astronauts left at Tranquility Base, containing words that will probably last for the life of the solar system: "Here men from the planet Earth first set foot upon the Moon, July 1969 A.D.," the engraved letters read. "We came in peace for all mankind." In Mission Control in Houston, a big screen bore these words: "'I believe that this nation should commit itself before the decade is out, of landing a man on the Moon and returning him safely to Earth'—John F. Kennedy, 25 May 1961." Another big screen bore these words: "Task accomplished—24 July 1969."[9]

CHAPTER 12

Sail Along, Silvery Moon: 1969–1970

We're all standing on the shoulders of Gene Shoemaker.
—LEE SILVER, *1999*

*Had Gene not been there, the Apollo crews would have
deployed the ALSEP and collected random samples. It would
not nearly have been the program we ended up with.*
—HARRISON SCHMITT, Apollo 17, *11 January 1999*

*If Gene hadn't devoted so much of his career to it, there
would not have been any geological studies on the Moon.*
—GORDON SWANN, *1999*

*Gene was the father of lunar geology, and the stepfather
of planetary geology.*
—STEPHEN DWORNIK, *1999*

BACK in the summer of 1968, the triumph of a moon landing was still a dream. That summer, Gene's long trip down the Colorado river gave him a chance to focus on what he was actually trying to accomplish as field geology leader for Project Apollo. Gene was not surprised to learn of the attempt to remove all science from the first mission, which took place while he was on the river. Even in the year preceding the launch of *Apollo 11*, Gene felt that maintaining the geology part of the mission would require an uphill fight. These negative feelings about *Apollo* began to surface in him in the year before the first Moon landing. He confided to Carolyn his frustration at the program's direction. The astronauts were being used not to explore the Moon's geology, he thought, but to act as maintenance men for their ships and instruments.

Sending a man to the Moon, Gene strongly believed, was useful only if that man could do what a machine could not do. The astronauts didn't have much time to spend on the lunar surface; Gene thought that the Apollo Lunar Surface Experiments Package (ALSEP) instruments should have been deployed automatically, freeing the astronauts for more exploration of this strange and beautiful world.

BLUE MOON

The summer of 1969 marked the last of a series of basic geology courses, a program some astronauts found boring. That August, Jack Schmitt, working hard to increase the scientific return in Project Apollo, paid a visit to Lee Silver to ask if he would do astronaut training. Silver spoke with Gene about the possibility and then suggested a field excursion with an astronaut crew. "If it works, it works," Silver mused. He met with the *Apollo 13* prime and backup crews, all of whom wanted to give the experiment a try in September 1969. Working with the prime and backup crews for *Apollo 13*, Silver spent a week in the Orocopia Mountains near Palm Springs, California. That September week was a big success and left everyone energized for an increased geologic role for *Apollo 13*. "Gene allowed me to take two Caltech vehicles. I brought the camping gear; we slept in the open, and I cooked for them. We ran long exercises. In the end the key man was Jim Lovell. He liked it." The result, which was made famous years later in an episode of Tom Hanks's television series *From the Earth to the Moon*, was a series of successful field trips. "The astronauts liked and admired Gene," Lee Silver remembers, "but most of them could not see how it was going to be relevant. After *Apollo 11*, the scientific implications of the voyage made a global splash. I came along at a better time than Gene did; the astronauts were much more ready to be persuaded. Gene was my friend and advisor at all points," Silver says. "The astronauts were my students, and I was a student of Gene's."

Less than a month after the field trip, Gene announced that he would leave Apollo. On October 8, 1969, a critical day in the lives of those who hoped to make a scientific mark on the Apollo program, Gene presented an informal talk on the status of Apollo. Gene had his audience recall President Kennedy's initial purpose in going to the Moon. The Moon was an unbelievable challenge at that time. Besides the engineering and technological challenge, the program was an inspiration to children from grade school through college and to many adults. Above all, Gene felt, Apollo should be an adventure in scientific discovery. In the preceding years Gene had tried to have astronauts chosen with the geological training to observe the lay of the land and to decide quickly and accurately which rocks were important to bring back.

Thinking that he was talking off the record to a friendly Caltech audience, he announced that within six months he would resign from his work with Project Apollo. Gene first attributed his reasons for leaving Apollo to his new appointment as chair of Caltech's geology department; he said he wanted to devote his full attention to this new role. But then he went on with his real motive, a sharp rebuke of NASA for considering the astronauts passengers on a trip to the Moon, their tasks largely limited to setting up and switching on experiments. He believed that NASA wanted to get the men there and back without any mishaps, as cheaply and as safely as possible. His conclusion: the space agency had no desire to improve Apollo so that the project would teach something about the Moon's geology.

Gene didn't know that a reporter was in the audience (although he did know that a published version of his remarks was available almost at the same time in *Engineering and Science Magazine*) and within a day Gene's caustic remarks were news around the country, and a surprise to colleagues who were working to increase the value of science in the Apollo missions and who were saddened by the loss of someone as enthusiastic as Gene. The press, however, overdid Gene's bitterness, Gordon Swann insists. He was not as bitter as the press led people to believe.[1]

If Gene didn't intend his spoken remarks that day to be generally known, he must have so intended his *published* remarks, for in the October 1969 issue of *Engineering and Science*, he painted a highly critical picture of NASA. The real reason for the start of the new space shuttle program, he wrote, "is that NASA wants to build big, new systems in space. NASA, after all, is primarily a big engineering organization, a good engineering organization, and good engineers like to build things." Yes indeed, and the result of what those engineers built allowed Neil Armstrong and Buzz Aldrin, Pete Conrad, and Alan Bean to walk on the Moon with *Apollo 11* and *Apollo 12* and collect rocks there under Gene's leadership. "NASA has just gone through an eight-year program of building a big system," Gene went on, "primarily for the sake of building a big system. This program is called Project Apollo." Gene slammed the Apollo designers because they tried too hard to keep President Kennedy's dream alive. To get to the Moon in time, they adopted the Lunar Orbital Rendezvous approach, in which the main craft would remain in orbit while a small piece of it, the LEM, would descend to the surface with only two men aboard, not enough to do a proper scientific program. "They are the consequences of focusing on getting a man to the moon and getting him back," Gene wrote, "rather than focusing on why he was going to the moon and what he was going to do after he got there."

In this article, Gene summarized his remarks at Caltech about what was wrong with ALSEP. Gene was disappointed that the men on *Apollo* were kept so busy deploying instruments that could have been set up just as well automatically. His hope was that the *Lunar* travelers would be trained and would function effectively as field geologists. We were the first generation, Gene believed, "to free itself from the gravity of Earth." The men of *Apollo*, he felt, should have used this freedom to act more as explorers. "If the astronauts were being used effectively," he wrote, "they would set out a package that would deploy itself . . . by remote control or automatically from the Lunar Module. In fact, the instruments deployed on the

Apollo 11 mission and the geophysical instruments planned for the succeeding *Apollo* flights could have been taken to the Moon more readily by an unmanned spacecraft. Had the instruments been ready, every one of them could have been taken to the Moon on a *Surveyor* spacecraft several years back."

It is interesting to read Gene's prescription for the remaining *Apollo* missions. Angry as NASA managers were with him, they ended up following his words almost to the letter: "The first order of business," he suggested, "should be to reorient the Apollo program to carry out an experiment in scientific exploration. The first returned samples have already shown that the lunar surface is rich in information about the early history of the solar system. In ten missions we should be able to find out whether a man can be used effectively to explore a planetary surface, or whether it would be better to send only unmanned instruments. If, at the end of these missions, the answer is clearly 'Yes, it was worth sending the men,' then, perhaps, we should start thinking about sending men to Mars. But let's be sure the program is planned to meet the purpose of exploration. I guarantee that it won't be worth sending a man to Mars just to demonstrate the technical feasibility of building a hundred-billion-dollar transportation system."[2]

Shine On, Harvest Moon

Actually, Gene did not really leave the Apollo program after his Caltech lecture, in spirit or even in person. "You could say," says Dwornik, "that NASA's reaction was one of disappointment, not anger, that Gene was leaving the program. 'How did we let this happen?' was a common theme." And despite his leaving the program physically, Gene did not lose his influence. Much as he might have tried, Dwornik remembers, "Gene never quite got out of being influential at NASA Headquarters." NASA scientists like Dwornik continued to rely on Gene's advice for *Apollo* and the planetary probes, particularly the Mars missions of *Mariner* and *Viking*.[3]

Several months after Gene's Caltech speech, the first lunar science conference took place. It was a major shindig, since it represented the work of hundreds of scientists. Since Gene was not working on samples, he did not present any paper. "But when several of us began to hear the way in which some of the observations were being interpreted," recalls Lee Silver," we all urged Gene to write the paper that appeared in the first set of Lunar Proceedings on Regolith Development. As reports came in Gene really was the man to put the picture together. It was superb and it had a profound influence."[4]

When *Apollo 17* thundered into the Florida night toward the Moon, Jack Schmitt and Gene Cernan were aboard, preparing for what would turn out to be a superb geological field excursion. Despite the confining schedule, they managed to make some astounding finds. As Schmitt explored the regolith of a crater he had named Shorty, he noticed some orange soil amid the unusually dark basaltic rock. Could it be oxidized rock, and could Shorty be a volcanic crater? More likely, an impact millions of years ago exposed a sheaf of orange and black glass that came from an eruption billions of years earlier.[5]

Had mission planners allowed extra time at the Shorty site, Schmitt and Cernan might have been able to collect enough samples to investigate its distribution. In any case, a hot discussion took place at Mission Control on whether it would be possible to extend the visit to Shorty crater. The decision was made that extending the visit so far from the LEM would be too risky. "Apollo provided outstanding science," Shoemaker complained, "but the scientific discovery came from the analysis of the samples brought back, not from the observations of the astronauts. And that was the issue. Why send human beings into space, if not to be prepared for the unexpected?"

Valid as this complaint may have been, in the light of history, Gene's words came across as a little harsh. The last three "J" missions (mission types were assigned letters based on their complexity), featured the Lunar Rover. This unique automobile was specifically intended to expand the area of geological exploration around each site, and the crews were specifically trained by geologists like Caltech's Lee Silver. Running a field trip from 240,000 miles away was not exactly field school for Geology 100 students. Although the astronauts did not spend as much time at Shorty crater as a field geologist would have liked, Schmidt and Cernan handled the situation expertly. As soon as one partner reported something highly unusual—like the red soil—they delayed collecting until they completed an in situ study of the material in its original, undisturbed setting.

It's also hard to ignore the imagination that a human presence brings to another world, no matter how they get there. Jack Schmitt proposed several names for the craters that he and Gene Cernan explored. Schmitt had named Shorty for a character ("Shorty" because he had lost his legs) in Richard Brautigan's *Trout Fishing in America*[6] though some mistakenly thought it was named for Shorty Powers, whose famous voice counted down the minutes and seconds to the Project Mercury launches. One V-shaped crater became Victory, in memory of Winston Churchill. Schmitt proposed "Jefferson-Lincoln Ridge" for a fault scarp, but that name was rejected as too political. "That list of names—Cochise, Sherlock, Brontë," Schmitt said, "represented names that I had grown to love as a kid and books that I was reading in the process of getting ready to fly to the Moon."[7]

Gene's dream for Project Apollo "was to try to create the opportunity to show what a well-trained human being could do, what kind of science he could do on the spot. This is not a kind of science that most practicing scientists understand because they don't do it. But in the six Moon landings, we never demonstrated that important discoveries could be made from field observation."

Gene never changed his mind about Apollo, despite the project's considerable scientific achievements. He never forgot his concern about the project managers wanting "to plot out every minute of every field traverse." "That plan," he argued, "mitigates against discovery." All Gene's training and experience as a field geologist led him to believe that the Apollo program, successful as it was, was a lost opportunity. "Discovery begins when you see something you don't expect," he said. "Then you stop and say, 'wait a minute, what is this? Maybe I'd better look over here.' Then you start out on a completely unplanned traverse to understand what you have seen. Apollo never provided that opportunity."[8]

Even if Gene's decision to leave robbed the program of his expertise, his mark had already helped set the direction for the remaining flights. Selling NASA on the idea that the Moon was an appropriate place to visit and study was one of Gene's greatest accomplishments. "Gene is enormously persuasive," Wilhelms wrote. "When he talks, everybody listens. I am told he was feisty and fiery before Addison's disease hit him in 1963, and I did hear a fine shouting match between him and [geologist] Henry Moore that year. Although he now has a calm, deliberate delivery, he is not at all boring even when he is talking about boring subjects. He has a hands-off management style and a way of making his listener feel that he or she is sharing in some grand project on an equal footing. Most of all he is passionately devoted to whatever project he currently has in view. The result was a generation of scientists convinced of the value of lunar geology and geologically-based lunar exploration."[9]

Considering the different minds and sets of enthusiasm that came to the Moon program, it seems amazing that so much science did get done. Although many scientists like Lee Silver contributed to the scientific energy, it might be said that Gene Shoemaker, who created the branch of science called astrogeology in 1962, and Jack Schmitt, who courageously fought for more hands-on geology and who personally saw to it during his 3 EVAs in *Apollo 17*, helped embody the spirit of the science project.

The Rille Not Taken

Davy Rille, a chain of small craters between the craters Davy and Alphonsus, is an example of the geology that Gene hoped to see from Apollo. The rille was a major interest of Gene's from before the Apollo program. It seemed possible at the time that these craters were the result of some volcanic event taking place deep below the crust, releasing its lava through a series of identical craters lined up like a chain at the surface. Davy Rille was one of several sites on the Moon, including a crater at the bottom of Alphonsus, and the Sulpicius Gallus Rille near the Mare Serenitatis, that were thought to be Maar volcanic vents, or diatremes, which are volcanic craters formed by sudden explosions of steam. There was a strong interest in Maar volcanoes as explosive vents for rocks deep within the Moon.

No one considered at the time that such a line of evenly spaced and sized craterlets could be impact in origin. If Davy Rille was truly volcanic, it should offer a rich sampling of material from deep beneath the Moon's surface. Accordingly, it became a prime landing site, and an Apollo mission was planned around it. In September 1970, the crew of *Apollo 15* visited a diatreme at Buell Park in Arizona. Had the Davy Rille been chosen as a site, this field trip would have been crucial. As it was, both Jack Schmitt and Lee Silver were familiar with the site and led a highly successful field excursion.

But decisions were being made virtually at the same time as that field trip was taking place. On September 2, *Apollo 20* was canceled, and *Apollo 15*'s H-style mission (an advanced field excursion but not yet with the Lunar Rover), which might well have been the visit to the Davy Rille, was also canceled. Davy was actually a candidate right up until the final choice was made for *Apollo 17* early in 1972; the final mission would visit the Taurus-Littrow valley at the southeast edge of the Mare Tranquilitatis.[10] With severe funding cuts axing missions *18*, *19*, and *20*, a full J-type mission for the rille, complete with the Lunar Rover surface transportation vehicle, was not about to happen.

It would have been fascinating, especially in terms of Gene's later career, had one of the *Apollos* landed at Davy Rille. Had the astronauts explored there, they would have seen the incredibly beautiful line of craters, most of them virtually the same size. However, the ancient lunar soil, brought to the surface by the supposed volcanic eruptions, probably would not have been there for the taking. In all likelihood, these craters are not volcanic in origin, even though at the time no other explanation made sense. It would have taken a line of more than a dozen comets or asteroids to strike the Moon, machine-gun style, to produce these results from impacts.

The study of the lunar rilles has long been controversial. While flying over volcanic features on Hawaii's Big Island, Gerard Kuiper and Robert Strom noticed partially collapsed lava tubes and channels that served as conduits for hot lava coming down from a crater. At the time, the only lava tubes that could be seen in Hawaii were the old ones, but a new Pu'u'o'o' cone has produced many active lave tubes and channels; my wife Wendee and I saw them while flying over the island. Kuiper and Strom noted the similarity between these structures and the sinuous lunar rilles, and they proposed that a similar process on the Moon formed these rilles.[11] However, Gene did not agree. He thought that these flows more closely resembled the suevite he had seen at the Ries in Germany, and he thought that a steady rain of material that had been thrown into the lunar sky as a result of the impact would explain these flow features.

But Davy Rille is different from the other sites. It is a chain of perfectly symmetrical craters arranged like linebackers on a football field. From information that was available before 1993, it would have been almost impossible to consider the origin of the chain as anything other than volcanic, whether from a maar-type series of steam explosions, or other activity along a rift.

Had a mission landed at Davy Rille, it may have set in motion this scenario: The astronauts would have found no solid evidence that these were volcanic craters, and they would have collected a barrel of samples. The origin of the chain of craters would have been a big puzzle—just as it was when other examples of crater chains were found on Jupiter's moons Callisto and Ganymede—until March 1993 and the discovery of Shoemaker-Levy 9. This comet, which appeared as a chain of twenty-one fragments, showed how the process of breakup and a subsequent chain-collision works.

What probably happened near Davy crater was that a billion years ago or so, a comet or asteroid passed within Earth's Roche limit and broke apart. (A planet's Roche limit is the distance from a planet within which tidal forces would cause a loosely formed body, like a comet, to disrupt. It was conceived by the mathematician Edouard Roche around 1850.) A day or so later, the pieces would have slammed into the Moon, creating the Davy Rille chain of craters within a few seconds. How interesting that a mystery Gene confronted in the early sixties was solved by his own observation three decades later! By the time Gene helped discover S-L 9, time had long since erased any bad feelings that some scientists and NASA managers may have entertained toward him. In fact, in that same discovery year—1994—geologist Don Wilhems summed up the feeling of many of Gene's colleagues in these words:

> Gene has been well recognized for his contributions and has now reached the enviable position of living where he wants [in a beautiful house in the woods near Flagstaff] and doing what he wants, when he wants, with only himself as boss. He deserves it. The science of lunar geology and I personally owe everything to this giant in the history of modern science.[12]

Should Gene have stayed with Apollo, despite his feelings? Considering that several fine geologists managed the geology experiments in *Apollos 14, 15, 16,* and *17,* Gene's decision to leave the program was a personal call that had much to recommend it. "I was PI (principal investigator) on *11, 12,* and *13,*" he concluded. "Thirteen never made it. I was worn down. I was just tired of fighting the battles, NASA, trying to get something done, pushing back. It was perfectly clear that most of the things I hoped could be brought about weren't going to happen."[13] Had he stayed, the program would have benefited from his wisdom, but since he did not stay, other geologists completed Apollo's marvelous contribution to our understanding of the Moon.

The post-Apollo planetary thrust, it also turned out, was not geological, but biological. The search for Martian life was a central goal for *Viking,* one of the most important post-Apollo missions. According to Steve Dwornik, the ratio of biology to geology papers and dissertations based on Viking was fifty to one, biology to geology. The modern discovery series probes, like Lunar Prospector, have returned to geology.[14]

"Apollo was not a highlight in my career by any means," Gene insisted. "The main issue for me was not flags and footprints, but to show why you needed a human being there." The proof of Apollo's failure in Shoemaker's mind comes from what happened afterward. "We stopped. Had Apollo really succeeded, we'd still be there exploring."[15]

CHAPTER 13

Chairman Gene: 1969–1972

From his cradle

He was a scholar, and a ripe and good one;

Exceeding wise, fair-spoken, and persuading;

Lofty and sour to them that lov'd him not,

But to those men that sought him, sweet as summer.

—SHAKESPEARE, Henry VIII, *1623*

How DOES a university professor teach a beginning geology course? When Gene Shoemaker was offered a position at his old alma mater as professor of geology and chairman of the Division of Geological Sciences, facing that challenge was one of the reasons he accepted so enthusiastically.

Gene actually rejoined the California Institute of Technology as a research associate in 1968. During the last years of the 1960s, virtually all of his time was spent, vicariously at least, on the Moon, with the exception of some field trips to Meteor Crater with Caltech undergrads. The chairmanship of the division, however, was not a task that he could do remotely. It would require his full attention, and his moving from Flagstaff to California at least while the Institute was in session.

The chairmanship was offered to Gene by an almost unheard of unanimous vote of the professorial staff of the geology division, a real honor from his alma mater and not an offer to be taken lightly. Since joining the Geological Survey in 1948, Gene rather missed the academic life, and early in the 1960s he considered leaving the USGS, and explored the possibility of a professorship at Berkeley.

With the pressure of Apollo, the prospect of returning to academia in 1969 was too great to resist. Moreover, Robert Sharp and Lee Silver helped ease Carolyn's reluctance to move to the Los Angeles area; Sharp sent pictures of the local area in an attempt to show that it rivals the Colorado Plateau in interest and beauty, and Silver stopped by at the Shoemaker's residence in Flagstaff with a very large bouquet of flowers.[1]

But as was usual in Gene's life, there were complications.

When Gene seriously began negotiating with Caltech, he expected that Apollo would be well underway by 1969, and that he could simply switch off his role in the space program and switch on his new role at Caltech. However, the January 1967 Apollo command-module fire set the whole schedule back far more seriously than Gene had figured, and he was faced with completing his work as principal investigator of Apollo's geological field investigations, *and* beginning his new Caltech position at the same time. Regardless of Apollo's schedule, on January 1, 1969, the beginning of Apollo's most historic year, Gene began his new work at Caltech, without yet moving to Pasadena.

Gene now had three somewhat incompatible jobs—Apollo, Caltech, and family. With his older daughter away at school and the rest of the family settled in Flagstaff, it made sense for the family to wait until the fall of 1969 to make the move to Caltech. Meanwhile, a vacation trip down the Green River through Lodore Canyon during that summer was precious time spent with Carolyn's brother Richard and his family, as well as with friends.

FAMILY LIFE AT THE SHOEMAKER HOME

The move to Pasadena was a critical time for the busy Shoemaker family, which had been brought up with a great amount of love but with a certain degree of strictness from Gene. All three children remember their father as very strict in some ways. They were not to talk with full mouths at the dinner table, and when the evening conversation got too enthusiastic, Gene would separate the children, a tactic that Linda recalls very seldom worked. "There were

20. A happy couple. Photo courtesy Pat Shoemaker and Paula Kempchinsky.

aspects of Dad's personality I didn't necessarily care for," Pat notes. Like most of his colleagues, his son found him impatient at times. "It was unncecessary impatience."[2] But at other times, when Gene might reasonably be expected to become terse, he was unusually patient. When he was teaching Patrick to drive one day, his son did not complete a turn quickly enough and bumped the car. "I sat there," Patrick remembers, "just waiting for the wrath to descend." But Gene looked at his horrified son, and laughed!

"We all sensed that Dad did not have the time to deal with kid problems," says Linda. "We could not be our usual goofy selves around Dad." Combined with his many absences from home, Gene's behavior was sometimes difficult for the children to understand. "Things had to be quieter when he was home," Linda explains. "I always felt we'd all get in trouble when he was home."[3] Even a certain specific look from her father was enough to let her know she had done something he didn't like.

The children sensed something else about their father, however. "Even if I was sometimes afraid of him," Linda says, "I always marvelled at how his mind worked." Christy recalls that her Dad was always teaching, whether a subject or an activity. "He'd lie on the floor and balance me," his oldest daughter remembers. He did allow his children to be independent. He let Linda live alone in Flagstaff so that she could continue her schooling while he and Carolyn were at Caltech. He had a natural trust in his children and it never occurred to him that they would disappoint him. "He knew I wasn't going to have parties or destroy his home in any way."[4]

"Dad would push you and push you and push you," noted Patrick's wife, Paula. "But in a way, that sharpened you. He would hone our rhetorical skills in our many conversations. Pat and I would be on one side of an argument, Dad on the other."[5] Linda's husband, Phred, has his own feelings. "Dad was like the godfather," he says. "Everyone could come to him with their problems and concerns. He was sometimes oblivious to what other people were feeling, but he did know how to say he was sorry."[6]

CHAIRING A DIVISION AT CALTECH

Becoming chairman of Caltech's Division of Geological and Planetary Sciences was a good thing for the division, says geologist Ron Greeley. Gene came at a perfect time, right at the height of the Apollo lunar landings and at the start of the planning for major planetary missions.[7]

"Gene really did not want to take on the job of chairman," his colleague Lee Silver remembers. "And anyone who has done a lot of things with Gene knows that while he was great on enthusiasm, he was not long on preparation." When he finally did arrive at Caltech as chairman, his enthusiasm and experience with Apollo charged the whole division. Seven months before the first human landing on the Moon, the division was alive with anticipation and excitement. If all went well, actual rocks from the Moon would be gracing the lab tables at Caltech. "A number of us were preparing to receive lunar samples," Silver recalls. "Some of my colleagues needed to know what Gene knew about lunar rocks. Simply analyzing the rocks out of context would not provide anything. There was a need to calibrate his time scale for the lunar surface."[8]

When the Shoemakers finally arrived in Pasadena, they found themselves in the largest house they had ever seen. The place at 312 South Holliston was near the heart of Pasadena, and only one block from the Caltech campus. It seemed intended for a division chairman; the house had a formal dining room and a rose garden.

The second half of 1969 was almost entirely devoted to the *Apollo 11* and *12* missions, even though the period between these missions was marked by Gene's famous announcement that he would leave the Apollo program and concentrate on his duties at Caltech. As we have already seen, Gene's October 8, 1969, speech was a shock to NASA but a tremendous relief to Gene and to the geological sciences division at Caltech, which had suffered with a leader who, until that day, could devote precious little time to his work there. As Gene settled in to Caltech, he found that in addition to teaching and Apollo, he had to concentrate on hiring, firing, and all the administrative details his new job required. By 1970 Gene felt he was well into the Caltech groove, and he found time to lecture at his old graduate school, Princeton, at Ohio State, and as far afield as the University of Barcelona in Spain. In 1971 Gene's schedule included time, once again, with the Apollo program. This time it was with the BBC: for *Apollo 14* in February and *Apollo 15* in July, he served on a panel that provided background to the British coverage of the missions.

At Caltech, Gene presided over a major expansion of the division. Now called the Division of Geological and Planetary Sciences, the program added a seismological lab and a large new building on campus that was appropriate for a geology program that emphasized the Earth as one of a series of worlds, all related and each one interesting. Yet all this time, he had his eye on doing a good job with the beginning geology course.

GEOLOGY 101 THE SHOEMAKER WAY

By what way is the light parted,

Or the east wind scattered upon the earth?

Who hath cleft a channel for the waterflood,

Or a way for the lighting of the thunder;

. . .

Canst thou bind the chains of the Pleiades,

Or loose the bands of Orion?[9]

—JOB 38:24–25, 31

As God queried Job out of the whirlwind, so Gene wanted to teach his students about the geological history of Earth: he would present a geological structure in the field, and then encourage his students to figure out how it got there by questioning them and by helping them paint the picture. One of Gene's proudest accomplishments as chairman was a revamping of the Geology 101 class to make it more field intensive. At the time, a typical first year geology course consisted of overviews of the two branches of geology, physical and historical. The physical side of geology looks at the process at work on the Earth; the historical branch explores how these processes have worked throughout the history of the planet. The typical texts sought to take advantage of the natural divisions between physical and historical geology in separate sections or even books. Caltech's geology department's course called Geology 101, had a greater emphasis on physical geology than on historical. For Gene, excitement about the Earth and its natural history was as normal as waking up on a sunny day, yet he was well aware that for many young students, finding an exciting career was a difficult process, especially in a generation beset by the political unrest of the time. Gene believed that for this challenging generation of intelligent and motivated students, a traditional geology course, with lectures, lab, and a few field trips would not be of interest.

Gene saw this as a major opportunity to make the course more strongly field oriented, with more of the teaching done out in the field and less in classroom and lab. The plan was rewarded in that more students from other departments began taking the course. Quite often Gene noted that the Caltech students had been so well trained to use their slide rules to the last decimal place that they ignored the overall picture. Gene often complained that students needed to understand basic things like significant figures. Gene believed in hands-on learning and that Nature itself was the best teacher; the more fieldwork, the better; as students saw for themselves how the Earth works, they gained a better understanding of it than they would in the classroom. Gene let his students see the Earth's geology at its best, at Meteor and Sunset craters, and the mighty Grand Canyon.

"Since there were not many students, we took one vehicle," says Larry Lebofsky, an astrophysics student who first met Gene in his Grand Canyon field trip. The weather was cooling off with the onset of winter, a steady snow was falling, and as they made their way into the canyon, "Gene showed lot of enthusiasm—after all, it was his crater! After the field day was over, we students slept on the floor at his Flagstaff home." In those days, the students were in awe of Gene. Even though he was a famous scientist, involved at that moment in the day-to-day decisions being made for the manned Moon landings, he spent warm hours with beginning geology students. "Gene could connect what was going on in space with what we saw at that crater on Earth."[10]

Of all the memories students have of Gene from those days, the best involved the field trips. The joke among graduate students: start a field trip with Gene—you supply everything, and he'll show up with his hammer. Throughout his life he would often show up only partly prepared, since he tried to do so much and was usually behind schedule. (Our observing field trips might have been an exception, for he was meticulous about being thoroughly prepared for them.) Always an optimist, he rarely brought rain gear; when it did rain, students would wear trash bags with holes torn out for head and arms. Gene was well aware that Caltech students tended to work late into the night and then sleep late, so he was not surprised when on field trips they would sleep on the bus. Then at each stop they would wake up, tumble out, listen to Gene's explanations, then climb back in and go to sleep again. Gene thought he had to make things interesting just to keep them awake! Carolyn Porco, now principal investigator of the imaging team for Cassini, a mission to Saturn, was a Caltech student in the later years of Gene's tenure there. She recalls the way Gene masterfully handled this story of how the different formations were deposited on the Colorado Plateau.

"As we caravanned from Pasadena to the Grand Canyon, we would stop here and there for Gene to explain the geological story. We stopped on a road with a view of the Grand Wash cliffs. Gene acted as a master of ceremonies as he explained the history of the river, working up a fabulous story that took place over millions of years. He turned the story into a nail-biting thriller, and then he concluded with: 'Now we are going to walk up to those cliffs and put our hands on the contacts!' We were spellbound! Gene was explaining something he knew and loved, and he made us all love it."[11]

To the students of the day, Gene taught by sheer enthusiasm. "Everything he did was his work," recalls Porco of the man whose time and energy went to geological problems from the time he awoke to his last sight of the day. When it came to organization, however, Gene usually left the details to others, or to no one. On overnight trips Gene would tend to stretch the day's excursion well into evening and suddenly realize he had no place to set up a camp for several truckloads of students. On at least one field trip the students were forced to catch some sleep while perched precariously on a slope!

As one of the trips neared its end, Gene's armada of rafts left the rush of Colorado River water and moved into the quiet waters of Lake Meade. In order to make their way across the lake, Gene brought a single outboard motor that he planned to attach to the last of the rafts, and then use the motor to push them across the lake to their vehicles. "He pulled on the string repeatedly," Porco recalls, "and couldn't get the thing to go! Turning red faced, he continued to pull on the string, getting madder and madder as the rest of us wondered how we'd ever get off the lake. Finally, he got the motor to start. His face broke out in a big smile and he laughed, as though it was all in a day's work! The group of rafts then happily motored across Lake Meade."[12]

Virtually everyone who took field trips with Gene, from students to alumni, had memories. One alumni trip was intended to be conducted at a reasonably relaxed pace, with the bus leaving Meteor Crater late in the afternoon to head back to Los Angeles and a stop for dinner in Sedona. But as the Sun sank lower, Gene had no intention of stopping. When he was reminded that the group was due in Sedona, a two-hour drive away, for a prearranged dinner, Gene was taken aback: "We don't need supper!" His own fervor prevented him from paying any attention to the comforts of people he was hosting.

"I think that Gene enjoyed toying with a group of victims that he could expect to come up with reasonable interpretations," remembers astronomer Donald Brownlee of a trip he took down Meteor Crater. "As we proceeded, Gene was alternately pointing out and asking questions—"what does this mean, what is this telling us?"[13]

Gene might not have been perfect as division chairman, but his popularity with the students was richly deserved. His open-door policy was genuine—students could literally march into his office and discuss the passions and problems of their Caltech careers at almost any time. It was in one of those impromptu sessions that Gene helped set the course of one well-known geologist, Joe Kirschvink, now director of Caltech's paleomagnetics lab. "I was a freshman at Caltech in biology," says Kirschvink, "and I just walked in. Gene dropped everything and talked with me for forty-five minutes. At the end of the forty-five minutes, he not only signed me up as a GPS [geology and planetary sciences major], but he put me under his grant for the USGS Palomag lab in Flagstaff! It was the best thing that ever happened to me."[14]

Of the many students who worked closely with Gene, Sue Kieffer's experience was certainly one of the most dramatic, and in a time when geology was essentially a male domain, unusual. Kieffer took her first year geology course at Caltech, and then took astrogeology from Gene three years before he became division chairman. She enjoyed the course, and its teacher, so much that she asked Gene if she could do a "proposition" for him in the summer of 1966.

Responding with his usual enthusiasm, Gene immediately suggested a mapping project at one of his favorite spots, the junction of the Green and Colorado rivers in Utah. Gene suspected that this area, a fault bed, might be a model in miniature for understanding the geologic history of the entire fault-block mountains and intervening basins that form the Basin and Range Province. Stretching through most of Nevada into the southern half of Arizona and Mexico and into the southwestern corner of New Mexico, this province is known for large, almost flat basins interspersed with high mountain ranges. The region was formed during the later part of the Tertiary period, when block faulting uplifted the mountains, a process that continued into the Pleistocene. One object was to see which of two fault lines was younger, the fault nearer the river or the one farther from the river.

"I had never hiked, never camped." Kieffer harks back to that memorable summer. "Gene had one day to take me to the Canyonlands. He parked me two hours from the nearest water, and four hours from the nearest town. We set up an eight-person tent, but Gene forgot the stakes for it." Gene returned to Flagstaff, where he was busy training astronauts, and left his student to return to Moab to buy stakes and set up her summer project. Although he did insist that she have a mobile phone, the local topography prevented it from working reliably. He did make a few suggestions, however, before he left. "Gene taught how the Navajo Indians could take a whole bath with just a cup of water," Kieffer recalls. "Put soap on your hands, then spit a little stream of water from your teeth!" Kieffer did complete her summer's work, and continued progressing through Caltech.[15]

Kieffer remained close to Gene over the years. "If you want to understand shock," Gene later advised her, "then you need to look at what happens to a real rock." To make that happen, Gene arranged a demonstration in the Coast Range Mountains east of Caltech. "Gene took me and his favorite rifle and found a boulder," Kieffer recounts. "He put a cardboard box over it in case any of the ejecta flew back at us. Then Gene said, 'Lie down, and shoot the rock!' 'What,' Gene said, 'haven't you fired a rifle before?' 'Of course not,' I answered. 'I'm a girl!' "

Under Gene's direction, Kieffer was about to learn to fire a rifle. "Lie down," Gene instructed, " and shoot the rock!" As the rifle fired, it hit her nose and broke her glasses. "The ejecta," Kieffer says, "came right back thru the cardboard box at us. So we had a perfect picture on the cardboard box of the ejecta coming out of that little crater we made." Despite the solo camping and rifle episodes, the two geologists remained good friends through the years, and as Kieffer's career evolved into volcanology, Gene took advantage of it. In 1979, he worked with Kieffer to interpret the newly discovered volcanos on Jupiter's moon Io. Ten years later, as *Voyager 2* sped past Neptune, Gene called again. "Sue, we have volcanos on Triton," he said typically, "here's a problem; let's go solve it!"[16]

ALL HIS FAULTS OBSERVED[17]

Although the students loved Gene as an inspiring teacher, neither Gene nor some of his colleagues were happy with his performance as head of the division. The minutiae of administration was not something he particularly enjoyed. Where his attention to detail was good for science, it slowed him down quite a bit when it came to running a department. His delays in handling administrative details could be costly; he was once accused of placing a large research grant in jeopardy by failing to respond to the granting agency before a critical deadline. The problem, as Caltech colleague Robert Sharp put it, was that Gene was very good at working with tremendous focus on a particular endeavor, often to the exclusion of everything else. However, if he began to lose interest in a project, he would spend so much less time on it that it would not get completed.[18]

If Gene's performance as chairman was questionable from the point of view of some of his colleagues, it was given almost universally high ratings from his students. "Gene was one of the most successful teachers we have had in our division," remembers Lee Silver. His immense enthusiasm and confidence inspired other people."[19] Gene was not a controlling person with his students. "I'll work over your first chapter," he advised Sue Kieffer on her thesis, "but then you're on your own." Gene was extraordinarily busy in those years, but he found the time to provide the guidance to get his students started. "Gene never had a lack of confidence in me, and never questioned whether I could succeed," says Kieffer. "Gene did not want us working in his lab on his grant. He wanted us to do what we wanted to do." When Kieffer offered to have Gene share authorship on a paper based on her thesis, he insisted that she enjoy sole authorship instead. "I'm a professor, Sue," he told his young colleague. "My job is to give away ideas."[20]

Generally, Gene's overcommitments were the subject more of teasing than criticism. Each year there was a banquet—the *Zilchbrau* event, made all the more fun as students and faculty did skits to tease each other. Gene would enjoy making fun of his own overcommitments as the man who had four hands. He would then do several tasks at once, like brush his teeth and talk over the phone, with Carolyn or son Patrick standing behind him and supplying two extra hands through his coat sleeves. The presentation was very effective; it really looked as though the busy division chairman was multifaceted. Gene particularly enjoyed those evenings when graduate students at Caltech had an annual chance to get even with their professors. "Gene chuckled and laughed," recalls Margaret Marsh—"his smile was so engaging."[21] Little memories stand out. For example his mere appearance on campus; where it would typically take a student rushing to class less than ten minutes to cross from one side of campus to the other, if Gene was involved, the walk stretched to half an hour as students in geology and other departments stopped, greeted him, and had conversations. "His rapport with students was really great," Sharp recollects, "he turned a ten-minute walk into a half-hour one, and a good memory."[22]

ALL'S WELL THAT ENDS WELL![23]

This was an exciting time in geology, but the demands of teaching and research often conflicted with Gene's need to keep Caltech's geological sciences division running smoothly. By 1972, Carolyn notes, "Gene was not satisfied with the job he did as chairman; administration was not something he enjoyed."[24] Gene decided to resign the chairmanship and looked forward to the free time he would have to continue his teaching, especially his field excursions, and his research. At this time, Gene rejoined the USGS and worked full-time with both Caltech and the Survey, and on alternating leave from each for half the year. This essentially meant that Gene had two major positions bidding for his time. By 1985 he left Caltech entirely, although he returned two years later as a visiting associate.

As his administrative obligations declined, Gene was freer to do the Caltech activities he enjoyed most, like his work in paleomagnetics and taking undergraduates on field trips to Meteor Crater and other favorite areas. It was on these strange and marvelous expeditions that Gene could return to the days he had loved as a young field geologist.

After one of those trips, late in the 1960s, Gene and a small group of Caltech undergraduates headed back toward the L.A. basin and Caltech. They stopped at a roadside restaurant for lunch and drinks. Except for Gene and one student, all the first-year students were under age. As they ordered their food and drinks, the waitress, dressed in Germanic costume as a mark of the restaurant's motif, looked at the young students and their leader. She accepted the orders for all the students, and then stared at the youthful soon-to-be chairman of Caltech's geological sciences division, the man who at that moment bore the responsibility for how the astronauts would explore the Moon.

"Sir, I need to see your ID," she demanded.[25]

CHAPTER 14

Shoot-out at the Moenkopi Corral:

1970–1972

Gene fought long and hard in areas not traditionally

related to geology.

—Lee Silver, 1999

Twisting and turning through the geological sciences, Gene Shoemaker's career helped bring distant scientific cousins, like planetary sciences, into the geological fold. In this chapter we look at how Gene's interest in the evolution of the stratigraphy of the Colorado Plateau, for instance, led him to the newly evolving field of paleomagnetism. "He followed that interest by establishing a paleomagnetism lab here at Caltech," says Silver, "but to do it he had to teach himself paleomagnetism. This was not a separate interest; Gene carried on this work in parallel to his impact research."[1]

Paleomagnetics: A New Area for Research

The promise of a new way of dating rocks was overdue. Geologists had relied solely on the natural decay of the radioactive atoms of elements like potassium in a rock to date it. This decay or change of unstable nuclei in the elements of a rock into other elements is constant over time and can be measured as a long-lasting clock. It is a good way to date rocks, but as the *only* way, it is subject to many types of error. Potassium decays to argon gas, for example, but argon can leak. The result is that the rock could be dated younger than it really is. The problem of accurate dating is central

21. Joe Kirschvink points to Gene's inscription on Mexican Hat formation, Utah. Gene enjoyed occasionally carving his name into places he shouldn't have, as he had as a child in his elementary school bathroom mirror.

to geology. Without it, geologists might have an idea of how old a particular rock formation is relative to another formation, but be unsure of its actual age. Since the early 1960s, when he established a geological time scale for major events on the Moon, Gene had been especially interested in geologic dating.[2]

The basic characteristics for magnetic dating of rocks have actually been known for thousands of years, as rock collectors have known that some minerals, rich in iron, have a magnetic field. But how this interesting property of some minerals can help scientists date rocks is a comparatively recent development. A rock acquires a magnetic field at the time it is formed. As sedimentation occurs, small mineral grains can settle and become magnetized in the Earth's magnetic tide. Finally, lava acquires magnetism as it cools. As rock becomes magnetized, it follows the orientation of the

22. Two of the greats: Walter Alvarez and Gene Shoemaker, 1993.

Earth's magnetic field and the position of the geomagnetic pole *at the time of its formation*, and it preserves that orientation as long as it is locked in place. This means that paleomagnetics—the study of how faint magnetic polarization of rocks can reflect the orientation of the Earth's magnetic field at the time of the rocks' formation—could be used to date rocks. In the late 1960s, this tool opened up the field of plate tectonics, where magnetic measurements of the ocean floor crustal rocks showed how the continental plates were drifting apart. This discovery was one of the most significant in recent geological history. Written into the submerged rock was every magnetic zone, every magnetic pole reversal, for the last 160 million years. Once this pattern was established, geologists could confirm it on other sediments around the world, like the Scaglia rossa limestone near Gubbio, Italy, which Bill Lowrie and Walter Alvarez were working on, and the sedimentary rock formation called the Moenkopi in northern Arizona, a special interest of Gene Shoemaker.

For Gene, the prospect of finding accurate dates for many of the formations he had studied in his beloved Colorado Plateau was invigorating, and as usual, he wanted to be in on the ground floor. He founded two paleomagnetics labs, one with his colleague Donald Elston, in 1972 in Flagstaff as part of the USGS, and a second at Caltech. The Caltech lab still prospers under the direction of Joe Kirschvink. Gene also began offering a Caltech course in paleomagnetics, with emphasis on the specialized nature of fieldwork. Field trips that emphasized paleomagnetics were different in that samples could not just be picked up, marked, and stored; they had to be scooped up from below the ground, their orientation preserved with the greatest care. The mere shaking of the drill bit is enough to shake up soft rock, allowing the magnetic particles to reorient themselves. For this purpose a heavy, hollow drill was used to bring up and protect the sample. The drill was powered by cumbersome McCulloch chain-saw motors, and its "bits" were actually one-inch tubes tipped with a quarter-inch rim of diamonds. To obtain a sample, a field geologist would drill a few inches into the rock bed by letting the diamond drill out the area around the sample. With the sample still attached to its bed, "staring at you like a bull's eye," according to Kirschvink, the worker would then mark its x, y, and z orientations using compasses. Once the sample was safely collected and returned to the lab, it would get inserted into a spinner magnetometer, a machine that would spin the sample to induce the rock's electromagnetic field and recreate the rock's field orientation in about thirty minutes. For hundreds of samples, such a procedure was slow and impractical.

In the early 1970s Gene learned of a new kind of magnetometer using superconducting components. He obtained funding and ordered one, and so the third superconductor magnetometer ever built was delivered to Caltech in 1973. Using liquid helium to keep its sensing coils at the required cold temperature, it could measure samples some sixty times faster than the old spinners could. "I still used Gene's magnetometer until 1997," says Joe Kirschvink of his long association with the laboratory. "In twenty-five continuous years it was used for more than two hundred papers. That machine was one of the most incredible investments of scientific research money I could think of." When the machine finally wore out, with leaking helium dewars, Kirschvink replaced it with a new one that he named for Gene.

The founding of the Flagstaff paleomagnetism lab dates back to 1968, in the middle of a hectic period. The Surveyor program was taking up much of Gene's time, and he was also fully occupied training astronauts at Meteor Crater, and at Sunset Crater in Hawaii, and in Iceland. He did manage to get funding for a traverse magnetometer staff to be transferred to this paleomagnetics lab. (Project Apollo hoped to determine whether the Moon has an iron core; it did not detect one but did find substantial records of a past magnetic field or fields.)[3]

The lab got its genesis in the Moenkopi, an early Triassic red bed in the Sinbad Valley some thirty miles north of Flagstaff. "We worked together for a few years on this project," geologist Don Elston remembers. Gene did the political work to fund the lab, and the two geologists worked both independently and together. "We had a lot of fun together, but we argued too much," Elston remembers, a sentiment echoed by other geologists about Gene's lively, but argumentative style. The two field geologists, however, were united in their enthusiasm to use the Sinbad Valley beds to understand the Earth's geomagnetic history at the time the sediments were laid down.[4]

Once the samples were collected, using them as a dating tool was as complex as it was exciting. It is common knowledge that the Earth's magnetic pole (now in Canada) is presently about eleven hundred miles from the true north pole at the top of the Earth's axis of rotation. Would not an ancient magnetic reading be inaccurate because of the difference between the two poles? Part of the reason for this difference is that the outer part of the Earth's iron core is fluid, causing the magnetic pole to wander. But by how much and over what time span? Paleomagnetic studies have shown that since the ice ages of the Pleistocene, the magnetic pole has wandered around the present geographic pole. In fact, for any period of about two thousand years, the *average* placement of the magnetic pole is approximately at the same place as the geographic one. Thus, when geologists take magnetic readings of ancient rock in specific periods of geologic time, they are confident that for any brief period of two millennia, the magnetic poles and geographic poles are at about the same spot.

If the magnetic pole can be directly related to the position of the true geographic pole, has that pole been steady through time? It turns out that it has not. With the evidence gathered from paleomagnetics, geologists have observed that although the geomagnetic pole as averaged out in 2,000-year intervals has been rather stationary for the past 30 million years or so, in the more distant past it has shifted remarkably. As far back as the start of the Proterozoic period some 2.5 billion years in the past, the North Pole was centered about where Arizona and California are now. Over time it wandered more than thirteen thousand miles in a vast semicircle past the Hawaiian Islands (a billion to 500 million years ago) then across the Pacific so that by the start of the Triassic some 190 million years ago it was over Japan. By the time of the extinction of the dinosaurs, the pole was approaching what is now the area occupied by the Arctic Ocean. This polar wandering does not take into account any effect of the drifting of the continents with respect to each other through geologic time. It turns out that on each major land mass the shifting of the geographical pole is different. Once these differences are adjusted for the inferred drift of the continents, they appear to coincide.

By the mid 1960s, paleomagnetics was becoming even more interesting because geologists were discovering that in the past, the Earth's magnetic field had actually reversed itself several times. Recording these reversals would further refine the paleomagnetic tool, allowing its use to become more exact. One problem with using magnetic field reversals in dating is that geologists could not use them for rocks older than Cenozoic era, at least not before two teams independently studied a limestone formation in Italy called Scaglia rossa. They saw that the Earth's magnetic field was reversing itself in rocks of that formation that were rich in specific fossils—the *foraminifera*—that flourished at the end of the Cretaceous period and were almost wiped out during the mass extinction that took place 65 million years ago. Thanks to this work, paleo-magnetic dating was now possible to a far older time than before.[5]

If the rocks were igneous in nature, they may have acquired mag-netism as they cooled; to detect this thermoremanent magnetism, the rocks cannot be exposed to daylight at any time from the time they were first covered until they are analyzed in the laboratory. Dating old lava was difficult because igneous rocks also recorded lightning strikes, but there was a way to identify the effect of the lightning overprint. Gene discovered that ancient lightning strikes could be used to provide relative dates for volcanic flows. Each time lightning strikes a specific rock, he reasoned, it remagnetizes it. A geologist can walk along an outcrop and with a Brunton com-pass note the number of lightning strikes per unit area. The more strikes, the older the rock.[6]

By the early 1970s some scientists saw the land beds as far more than just a confirmation of ocean floor data, which was believed to be cleaner and less disrupted with the contaminants of weather-ing and erosion. Gene helped pioneer the idea of using sediments now on land to unravel the pattern of paleomagnetism much fur-ther back in time than 160 million years. Using the tools then avail-able, Gene mounted field expeditions, collected samples, and ran tests on those samples.

Controversy on the Moenkopi

With the superconductor magnetometer Gene and his team were able to get reliable magnetic grid patterns back to the middle of the Paleozoic era some 240 million years ago. Gene was adding some 80 million years to our understanding of the geomagnetic history of the Earth by suggesting that the Moenkopi formation had faithfully preserved the Earth's magnetic pole reversals. "Start-ing at the bottom of the formation," remembers Joe Kirschvink, "you get one polarity, and after a quarter meter or so of beds, bang! you're at another polarity one hundred eighty degrees away from the lower one. Moreover—and this was Gene's big contribution— you can trace these from butte to butte and from section to section, and you can tie the magnetic reversals to the details of the stratigraphy." Thus, in this formation, ancient reversals of the Earth's magnetic poles were well recorded.[7]

One of Gene's students, Michael Purucker, worked with Gene on a particular magnetic reversal that could be traced from outcrop to outcrop, from New Mexico through to California. Using evidence like this, Gene proposed that all these beds, especially the Moenkopi, showed evidence of primary magnetism—that the magnetic field in these rocks was unchanged since they were originally formed, and that they reliably indicate the direction of the Earth's magnetic pole near the time they were laid down.

Gene saw the Moenkopi as a litmus test for his idea that sediments on land had sufficiently good magnetic records to bring back the paleomagnetic record to an earlier time than could be read from the ocean floor. But many other geologists, particularly Ed Larson and Ted Walker, did not agree with this interpretation, contending instead that the red beds they studied in Baja California had been remagnetized as they reddened, to the point that one could not do useful magnetostratigraphy on them. The controversy erupted at several geology confabs and grew so bitter and divisive that scientists on a National Science Foundation review panel began denying funding to geologists wishing to do magnetostratigraphy on these beds.

The final battle, at the 1976 Denver meeting of the Geological Society of America, has become famous as the Shootout at the Moenkopi Corral. Larson and Walker suggested that all these beds, including those of the Moenkopi, were magnetized over the course of millions of years and not, as Gene maintained, shortly after deposition. While Gene had to admit that not all Moenkopi beds yielded perfect magnetostratigraphy, plenty did. The shootout was rancorous and bitter, with students on either side hanging onto their proponent's every word. "I've never seen bloodletting like that," says Don Elston.[8]

Using evidence from Baja California rocks nowhere near the Moenkopi, the naysayers insisted that the Moenkopi beds could not deliver reliable magnetostratigraphy. The debate was moderated by geologist Robin Brett, who, in the end handed the argument to Shoemaker's team. He agreed that the evidence confirmed that the Moenkopi beds reflected the true state of the Earth's magnetic pole at the time of their deposition.

Despite his proactive role in the red-bed controversy, Gene actually published little of his research in paleomagnetism. "Gene was not too good at following up on the publications," says Kirschvink, echoing others who made similar comments about his other work, particularly that from Australia. Gene's mind was so rich with ideas that he often did not have the time, or make the time, to publish the details of all of them.[9] But had Gene made the time to publish everything in his primary field, all the other field contributions of his multifaceted career, like the paleomatics of the Moenkopi, would probably never have come to pass.

The Little Prince Revisited: 1972–1979

I spent the first half of my life wanting to go away to the
Moon, and the second half of my life wanting the Moon
to go away.

—GENE SHOEMAKER, *circa 1985*

WHEN Antoine de Saint Exupéry published, in 1943, his allegory about a prince who lives on an asteroid, Gene Shoemaker's mind was still five years away from starting to dream about the Moon.[1] The beginning of Gene's search for asteroids and comets did not occur, as some believe, as a sideline after Gene formally left Project Apollo. Its roots go back to that morning in 1948 when Gene first looked at the Moon and wondered about its features, thence to a seminal 1963 paper on "Interplanetary Correlation of Geologic Time." "The code in which these planetary histories are recorded," he wrote, "consists of bodies of rock and rock debris. This code will be cracked by geologic mapping, for it is the spatial relationship of different bodies of rock that tells the sequence of events."

Even this early in Gene's career, one can sense the direction it was to take. "The interaction of the the solid material in the solar system by collision . . . provides in principle two independent methods for the interplanetary correlation of geologic time. First, if the frequency of meteoroid impact and its variation with time on the different planets can be established, the age of rock bodies exposed on their surfaces can be estimated from the distribution of superimposed impact craters."

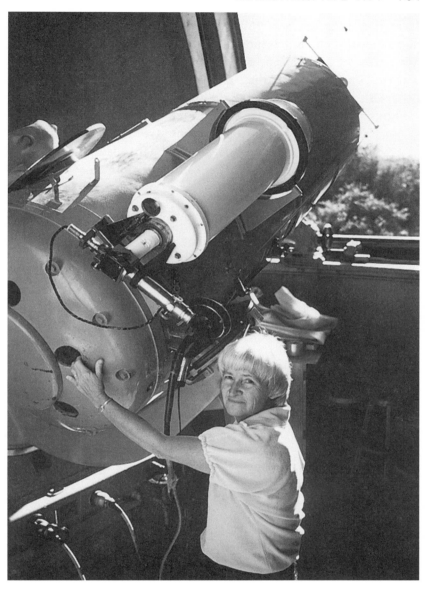

23. Carolyn at the eighteen-inch telescope. Terence Dickinson photograph.

In examining the roots of this important period of Gene's life, this paper is seminal. Coming out in a year of poor physical health, this paper set a direction for the young geologist's career. The extent of Gene's thinking then about the environment of the solar system through geologic time is remarkable. The paper's ending contains an even more intriguing thought, one that anticipated the discovery of meteorites blasted out from from Mars by a quarter of a century: "A second potential method depends upon the transport of impact debris from the moon and other planets to the Earth, where the debris becomes incorporated in the terrestrial stratigraphic record.

"If some small amount of material escapes from Mars from time to time," the paper concluded prophetically, "it seems likely that at least some very small fraction of this material would ultimately collide with the earth. Whether it could ever be recognized is difficult to say, but the possibility that such material could carry organic hitchhikers, however remote, may present a vexing question to those who are concerned with the origin of life."[2]

Gene spent much of the next ten years with Ranger, Surveyor, and Apollo. Despite his disappointment with the lack of science in Apollo, the projects did allow him to refine his thinking in regard to asteroids and comets. In 1972 Gene looked skyward in a different way, not now toward the Moon but away from it. Instead of studying the results of impacts, Gene wanted to study the comets and asteroids themselves. What sort of orbits were these asteroids following that could bring them so close to planets? How numerous were they? What was their size? How were they composed? On this new venture, Gene would go into the dark, moonless sky and search for the telltale trails of asteroids and comets that could someday strike the Earth. To find and study these asteroids, dark sky conditions must prevail. And so, rather than welcoming the Moon, Gene indeed wanted it to vanish.

When Gene resigned from the chairmanship of geology at Caltech in 1972, he looked forward to indulging in other research. He had two things in mind: One was the founding, with Don Elston, of Flagstaff's paleomagnetics laboratory. The other was a search for asteroids and comets whose orbits made collisions with the major planets a possibility. The following year he founded the Palomar Planet-Crossing Asteroid Survey.

AN ASTEROID SEARCH

As of mid-1999, the Minor Planet Center's Cambridge, Massachusetts, file lists some 11,500 asteroids whose orbits are well known, plus at least that many in need of more observation. Most of these objects are in the large "main belt" that lies between Mars and Jupiter. There are others whose orbits have, in their long history of orbiting around the Sun, altered their orbits so that they intersect the orbits of the inner planets. A small group of asteroids, the Amors, travel so far inside Mars's path that they approach but do not quite reach the orbit of the Earth and can briefly make close passes by us. Still other asteroids, called Apollos, actually venture across the Earth's orbit, and a few, called Atens, orbit even closer to the Sun in periods of less than a year. Gene wanted to know how often asteroids, and their cousins, the comets, have struck the Earth. In 1959, while Gene was in the midst of his research with Meteor Crater, only nine Earth-crossing asteroids had been found, and these were found incidentally as part of other search programs. By 1973 a few more had been identified. These asteroids are at least a kilometer or larger in diameter, asteroids capable of inflicting global damage should they strike the Earth, a concept that would not reach scientific consciousness until the end of that decade.

By the time humanity made its first landings on the Moon, very little was known about the few Earth-crossing asteroids that had been discovered. One of them was even utterly lost. On the night before Halloween in 1937 (and exactly a year before Orson Welles's Fireside Theater production of *War of the Worlds*), an Earth-crossing asteroid raced to about twice the distance of the Moon. It has not been seen since. But that asteroid, named Hermes, will be back some day in the distant future, either to make another close pass by the Earth or even to hit us. The collision would cause far greater damage than the asteroid that struck northern Arizona to form Gene's cherished Meteor Crater some fifty thousand years ago.

Gene knew that in order for a statistical history of the Earth's encounters with these bodies to make sense, many more Earth-approaching asteroids had to be found. As Gene prepared to start preliminary work in 1969, he met with Eleanor Helin, a young Caltech scientist who already had a serious interest in the subject. She came to Gene with some technical experience, having worked with Harrison Brown and Bruce Murray on lunar geochemistry. "I was primed to go and try to understand the flux of asteroids passing into the vicinity of our Earth/Moon system," Helin recalls, "so we were both prepared to join forces to study what I thought could be an important interrelationship between asteroids, meteorites, and craters."[3]

Since Gene had known Eleanor Helin for some time, he agreed it would be a good idea for them to work together, and he encouraged her just as he encouraged all his colleagues and students. Their first step was an investigation of the population flux of planet-crossing asteroids. Concerned that his crater studies had not provided sufficient data to give him good population estimates of asteroids and comets that could collide with planets, he asked Helin to "to track down every scrap of information on all the Apollo asteroids that had been discovered up to that time."[4] Helin obtained copies of the original-discovery photographic plates from places as far apart as West Germany, Belgium, France, and South Africa, and as far apart in time as two thirds of a century. She did this work enthusiastically, even though there were no funds for her efforts for the first six months.

The Heidelberg photos were wide-angle camera images dating from the early twentieth century, taken by astronomer Max Wolf. The early observer George Van Biesbroeck worked in Belgium before he came to America during the First World War. More recent photos were taken at Harvard's Boyden station in Blomfontein, South Africa, by Bart Bok, the famous Milky Way astronomer, and by J. S. Paraskevopoulos, who observed there for many years. Bit by bit, from these old plates, Helin gathered information about what these objects looked like when they were first spotted on long-exposure photographs. She then compared them with later images taken at Palomar's forty-eight-inch Schmidt camera for its first sky survey.

Using a twist on the old geological adage, "the present is the key to the past," which we first explored in chapter 5, Gene calculated the ancient collision rate based on the number of craters he had already investigated on the Moon. In this case, the past was the key to the present. The lunar crater record, however, did not answer questions about the nature of the objects that hit, nor did the Moon hint at the orbits of these objects. The information that Helin obtained from old photographs was the beginning of a real understanding of interplanetary correlation through geologic time, but the number of asteroids was simply not sufficient. What Gene was looking for could be obtained not from impact craters but from observing the asteroids and comets themselves. Gene wanted to date planetary surfaces. Lacking rocks on these surfaces to date (at least before Apollo), the next best thing to do to learn the age was to count the surface density of craters, determine the rate at which asteroidal and cometary projectiles strike, and multiply. Gene realized that although spacecraft were obtaining data about the crater density, there was a big gap remaining in our understanding of the strike rate.

In discussing this problem, Helin urged Gene to consider going out so that they could take their own photographs and begin a search themselves. The eighteen-inch telescope available there as a part of Caltech, seemed an ideal instrument for this search. As principal investigator for PCAS, Gene planned to apply for telescope time and conduct the scientific inquiry, but at this early stage he doubted he'd do much direct observing. It turned out that Caltech's Palomar Observatory, home of the 200-inch reflector, then the world's largest telescope, had a small eighteen-inch-diameter Schmidt telescope that was seeing only occasional use. Gene jumped at the prospect of putting this beautiful instrument to regular use.

The Eighteen-Inch Telescope

With Gene already an established figure at Caltech, the fact that the institute already owned a telescope that was ideal for his research was a remarkable coincidence. As Palomar's first telescope, the eighteen-inch has a noble history. It was designed by Russell Porter, an Arctic explorer, artist, and amateur astronomer. In 1925 Porter launched an annual telescope-makers conference on Breezy Hill near Springfield, Vermont. Porter called the site "Stellar Fane," for Shrine to the Stars, and the name was later compressed to Stellafane.[5] In November 1925 a *Scientific American* article started a sensation by showing how amateur skywatchers could build their own telescopes as a way of beating the high prices of commercial refractors.[6]

By 1930 Porter's reputation had spread across the country and attracted the attention of George Ellery Hale, who was completing the design for a 200-inch reflector telescope that would be the largest on Earth. After Porter moved across the country, he began drawing blueprints for the 200-inch. In 1936 he designed the eighteen-inch Schmidt camera, the first telescope permanently mounted at what would become one of the world's most famous observatories. Glad to return to "pushing glass," Porter completed the optical figuring of the new eighteen-inch telescope, allowing it to begin its task of searching for exploding stars, or supernovae, in distant galaxies.[7]

The eighteen-inch Schmidt is a photographic telescope capable of taking pictures of large areas of the sky during each exposure. Since each film exposed through this telescope covers 8 3/4 degrees of sky, the equivalent of more than seventeen moons lined up, the instrument is ideal to use in a patrol program for wandering asteroids and comets. (The eighteen refers to the diameter not of the telescope's mirror but to the lens at the front of the tube; the mirror inside is considerably larger.) For the new PCAS program, the telescope was about to embark on a new career.

OBSERVING BEGINS

In 1972 the Shoemaker-Helin team drew up a proposal to see what would be involved in a search for asteroids in the vicinity of the Earth. Having explored lunar craters to get an idea of the number of impacts in the past, it was now time to search space to glean how many potential impactors there were at present. "Glo deserves a lot of credit," Gene contended, using Eleanor Helin's informal nickname, "for pushing enthusiastically to start this program at the beginning."[8]

"After exploring the possible available telescopes at Palomar," Helin says, "I made an appointment with the director, Dr. Horace Babcock, and explained to him what I wanted to do using the eighteen-inch Palomar Schmidt."[9] Babcock thought that Tom Gehrels's Palomar-Leiden survey for asteroids, completed more than a decade earlier, had accomplished the same task. However when the proposal was submitted, the program was granted several nights per month on Palomar's eighteen-inch Schmidt.

At the time, Gene suspected that about two thousand asteroids the size of Hermes (two kilometers or larger) could be in orbits that could cross that of the Earth's and hence be a threat to our planet, each one packing the wallop of a multimegaton bomb. Gene calculated that by photographing 250 different fields of the sky each year, they would perhaps find some four collision candidates.

At this early stage in the program, the plan was to take, using filters, a twenty-minute exposure followed by a ten-minute follow up shot. A fast-moving asteroid would then leave its trailed signature as it moved across the field. "There was a lot of sweat and tears for each observing run," Helin says, recalling those early days when it was difficult just to get some 1,300-square degrees of sky photographed each year. And getting the telescope working was only part of the problem. "Roadblocks appeared on a regular basis," she remembers. "I'm sure the staff at Palomar didn't think I would last a year."[10] Helin was the first woman observer permitted to regularly use the lodgings provided at Palomar Observatory's dormitory building. Dubbed the Monastery, the residence was a male enclave until Helin broke the gender barrier in 1973. During her three nights per month at the eighteen-inch, Helin averaged a take of seven pairs of photographic films per night. On Independence Day in July 1973, only six months after the project had begun, she found her first new Earth-crossing asteroid. It is now numbered as 5496. Encouraged by this first find, Helin pressed on, aided mostly by students and occasionally Gene for her three-night observing runs. But that new Apollo was just a teaser; the return was rather sparse during the years that followed.

When the NASA grant expired, a few months passed before new NASA funds were in place for the project, during which time Helin continued the observing work without pay. Once funds were restored, she also found students to assist at the observatory. Meanwhile, concerned that the discovery images alone were not enough to provide an understanding of an asteroid's physical characteristics, Gene proposed a program on another telescope at Palomar, the sixty-inch Oscar Meyer telescope southeast of the eighteen-inch, to make follow-up observations of objects discovered with the smaller instrument. If this proposal hadn't been turned down, it would have pioneered the field of the physical study of asteroids. A decade later I was part of a group of observers at Tucson's Planetary Science Institute that conducted a related study. In our multiyear project at Kitt Peak, we deduced sizes and shapes of several large asteroids from their light curves.[11]

By the program's fifth anniversary in 1978, Gene suspected that most of his supposed two thousand near-Earth asteroids were closer to one kilometer in diameter, about half the size he had expected. With Helin, he published a report the following year in the planetary science journal *Icarus* that reviewed the progress the team had made and its finds of twelve new planet-crossing asteroids.[12] Despite the slow progress, the team doubled the count of Earth-crossing asteroids observed over the last century in just five years.

It would be harder to find asteroids that were close to the limit of the eighteen-inch camera, and Gene thought of applying for time on Palomar's larger Schmidt camera, the mighty forty-eight-inch, which was used some twenty years earlier in the Palomar Observatory Sky Survey, when it photographed the entire northern hemisphere heavens. How could one get more out of the eighteen-inch telescope? An idea to substantially improve the detection rate was crossing Gene's mind, an idea that would use the same stereographic principle as that employed for some of the stereo pictures *Surveyor* and the *Apollo* orbiters had taken of the Moon and system geologists use with aerial photos to produce geologic maps. These pictures gave the features a perception of depth. If observers took short exposures and then examined the films in pairs, the moving asteroids would appear not as trails but as single images standing above the starry background.

It turned out that Gene was not the first to use a stereomicroscope for discovering asteroids and comets; Max Wolf had used one at the beginning of the twentieth century in Heidelberg. However, he had difficulty seeing in stereo. Wolf then switched to a blink comparator, a device that worked differently in that the observer would look alternately and rapidly at one image, then the other. This is the type of instrument that Clyde Tombaugh used in his sky survey from 1929 to 1945, and with which he discovered the ninth planet, Pluto, in 1930.[13] Half a century later, Gene was about to design a new stereomicroscope. In 1978 Helin located a company called McBain Instruments, and she, Gene, and company representatives met at the eighteen-inch telescope to consider a design for a stereomicroscope capable of holding two six-inch-diameter films at one time. The result was a stereomicroscope—a tool for discovery—for $3,571.[14]

In May and June of 1980, Gene conducted the first tentative observing runs using the stereomicroscope, with Helin at the eighteen-inch. The early results were encouraging, at least to Gene. By comparing the brightnesses of new discoveries with those of the known objects, Gene calculated that they were finding asteroids as faint as eighteenth magnitude, which would suggest that these faint asteroids were about a kilometer in diameter. "We were coming back from every dark run with lots of new asteroids," he noted, "which was quite a happy contrast to what had happened before."

24. Gene and Carolyn. National Geographic Society photo.

Carolyn, too, was beginning to take up an interest in this program that year. Christy, Pat, and Linda were grown and gone from home, and Carolyn longed for an activity that would add as much motivation to her life as geology did for Gene's. For a year in the late 1970s, she worked in a florist's shop—fun, she recalls, but not something she wanted to do for the rest of her life. In 1980 Gene suggested that she might like to try her hand at the asteroid project. One of his students, Bobby Bus, had just returned with a series of large glass plates he had exposed at Australia's Siding Spring Observatory, and Gene suggested that Bobby Bus teach Carolyn to look for the trailed images of asteroids on his photographs.

Carolyn quickly developed a facility for this work, and her new interest excited her family. "When Mother decided to join Gene in his work," son Patrick says, "that was great. It really hit the spot for both of them." At first, the discovery rate for asteroids on those big plates was very slow. "I had no interest in observing," Carolyn says, recalling her early months with the program. "Being a morning person I did not know how I could ever stay awake at night!" So through the end of 1981, Carolyn's role in the program was limited to the lowest sub-basement in Caltech's astronomy building, searching the Australian plates for asteroids, and measuring the asteroids on those plates with a view to extracting exact positions. The task of measuring the position of an asteroid on a photographic plate is complex, requiring an asteroid's position on the X and Y axes to be compared with the placements of several surrounding stars. The early attempts at measuring asteroid positions were very difficult, with any number of possibilities for error.

Helin used the new stereomicroscope during her summer observing runs. "I used it, but it did distort the muscles in the eyes and was tiring for long-term use."[15] Although she at first continued to use the old microscope, which could examine only one film at a time, she eventually switched over to the new stereomicroscope and used it successfully for the next sixteen years. Meanwhile, Gene asked Carolyn to try scanning some films. "I found that I had no trouble with it"—Carolyn describes her initial rush of excitement in using it—"and when Gene saw how much I was enjoying it, he decided to leave all the scanning to me."

Using the stereomicroscope gave quite a boost to the project, and in 1980 a second was ordered.[16] Asteroids were cropping up quite frequently on the films. In January 1982 Gene, suspecting that Helin was losing interest in the small aperture of the eighteen-inch, experimented with a new program that used both the eighteen-inch and its larger sister, the forty-eight-inch Schmidt camera. In addition to the survey photographs, he and Helin would also take plates of specific regions, on the large forty-eight-inch Schmidt, in hopes of recovering periodic comets on their way back to the inner solar system.

Carolyn was getting interested in observing as well. Bobby Bus and Quinn Passey, both experienced observers on the eighteen-inch, taught Carolyn how to load films into that telescope, point it, guide the telescope so it would precisely follow the sky, and develop the films. While Helin observed with Gene at the forty-eight-inch, Carolyn would conduct the eighteen-inch program, and soon she was proficient enough to use the telescope without assistance.

Because the eruption of the Mexican volcano El Chichon in the spring of 1982 spilled tons of fine particles of sulfuric acid into the upper atmosphere, the sky over Palomar suffered during much of that year. Helin did discover an Apollo asteroid in February designated 1982 DB: now named (4660) Nereus. "A benevolent seagod associated with ancient origins (mythology, if not science)," the citation read, this deity "had the power of prophecy."[17] But if Nereus was prophesying anything that February, it was a change in the program. By the fall of that year, friction between Helin and Shoemaker had become serious. One cloudy night at the forty-eight-inch, the two astronomers had a discussion that evolved into a bitter argument. Helin left the forty-eight-inch telescope dome and hurried back, with the only key, to the Monastery, Palomar's dormitory. Gene closed up the telescope, but when he reached the Monastery door, which was never locked, he found it locked! No one seemed awake on that cloudy night, including Carolyn, who had already ended her night at the eighteen-inch. Gene grabbed some small stones and started throwing them at the window. This exercise didn't wake Carolyn, so Gene started walking the considerable distance toward the sixty-inch telescope, hoping to find someone there. Before Gene got to the telescope, he encountered one of Palomar's night assistants, who was happy to unlock the Monastery door at last.

That was the last straw. Gene decided to end the Palomar Planet-Crossing Asteroid Survey, at least as a project with his participation. He hoped that the split would be better for him, better for Helin, and better for science. Helin disagreed. "Gene can be charming." She explained how she thought he would do anything to get his way. "But then he will waltz in and say that these are his programs and his results."[18] Other observers, including several who observed with him in later years and who worked with him in earlier projects, questioned that assertion. Gene wrote the proposals and reports and oversaw the work in his programs. He seemed to go out of his way, they would argue, to make sure that both colleagues and students working with him received the proper credit for their contributions.

When Gene decided that the program would split, he did two things that he thought would please Helin but, it turned out, only annoyed her. One was that he gave her all the funding and resources of the Palomar Planet-Crossing Asteroid Survey. However, Gene was now teaching only a quarter-time, so he no longer found it necessary to remain on the Caltech campus. He and Carolyn had even relocated back to their Flagstaff home. Since Gene was no longer at Caltech, it was not possible for Helin to conduct her program from there. Transferring her project to Jet Propulsion Lab seemed to Gene to be a solution. He also sought funding for a completely separate program, the Palomar Asteroid and Comet Survey (PACS), for him and Carolyn. In the new program he had some new techniques—like the use of faster film, and filterless exposures—that he wanted to try. He also thought that Helin was no longer interested in the eighteen-inch telescope since she could work on the larger, forty-eight-inch Schmidt. "Gene tried to define his program as a different one," Helin recalled bitterly, "but it was not different at all." After the split, Helin continued to build up her program with both telescopes, and its successes were later the subject of a *Time* magazine article.[19] Her work at the eighteen-inch continued until the end of 1994, during which time her team discovered fifteen comets and many Earth-approaching asteroids.

THE PROGRAM

As the new program evolved, exposures shortened considerably, allowing time for vastly more sky coverage. Gene and Carolyn also took advantage of the devlopment of the new fine-grained Kodak Technical Pan 2415 film (later the thicker-based 4415), and also a method to speed up the film emusion—hypersensitizing—that revolutionized the program. During the course of four or five clear nights of winter observing, when the night can exceed thirteen hours in length, Gene and Carolyn, assisted later by Henry Holt and later still by me, might expose three to four hundred pieces of film covering much of the available sky. We typically took four eight-minute exposures of different areas of sky, then repeated the exposures so that each area was shot twice with about a 45-minute break in between. Carolyn then scanned the films in her stereomicroscope. Each film in a pair would be identical with its mate, except for any moving asteroids or comets that would appear to float atop the surrounding background of stars. We thought this was a most elegant way to search for asteroids and comets.

"I brought the tools of the geologist," Gene later said, "to the task of looking for rocks in the sky."[20] But as resources for planetary science diminished during the 1980s and early 1990s, both his program and Helin's were assured a difficult time for funding. It turned out that Gene and Carolyn's PACS suffered most from this. Over the years, people who assisted the Shoemakers, including me, had to work only for the love of it. The only exceptions were occasional small grants for students, but not a nickel was available for the observers' time. To get what funds he could, Gene had to bootleg, as he put it, the money to cover the cost of film, transportation, and housing. It was a dismal way to have to run a program that was destined to cap Gene's career with a remarkable discovery.

CHAPTER 16

A Ship Sails: 1977–1989

Whither I go, thither shall you go too;

To-day will I set forth, tomorrow you.

—SHAKESPEARE, Henry IV, *1596–97*

Gene had a whole new approach to planetary geology.

He combined observations of planetary surfaces, from

spacecraft or telescopically, with field work on terres-

trial analogs, numerical studies, and laboratory simula-

tion. He was never locked into a point of view. We all

use this approach in the community now.

—RONALD GREELEY, 27 July 1999

AT THE SAME TIME as Gene was ramping up his survey of small, nearby bodies of the solar system, an ambitious group of scientists was preparing two spacecraft, each the size of a room, for an odyssey quite unlike any other in the history of humanity. If all went well, these craft would set forth to planets and moons in the outer solar system. The project was named Voyager, and the craft were taking advantage of a unique planetary lineup involving Jupiter, Saturn, Uranus, and Neptune, which takes place once every 176 years. The journey was planned to allow a single craft to visit first Jupiter, and then use Jupiter's gravity to give it the required velocity increase to send it to Saturn. As it passed the ringed planet, it would in turn use its gravity to get the velocity increase to send it to Uranus, and finally use the pull of Uranus to hurl it forward to Neptune.

The original plan for this complete "Grand Tour" was cut from NASA's budget because of conflicts with the development of the Space Shuttle. The Grand Tour was demoted to two missions called MJS (Mariner Jupiter-Saturn) and MJU (Mariner Jupiter-Uranus). The latter was cancelled, and MJS was renamed Voyager. The new pair of craft would visit Jupiter, Saturn, and Saturn's giant moon Titan. Uranus and Neptune were not in the original or official plans of the MJS program. However, mission planners and engineers could not accept that these marvelous craft would travel through space unable to take advantage of the lineup. They added something rather new to the craft's computer, a program giving the computer some autonomy in selecting new targets while in flight. The intention was that the craft would be sent to Jupiter and Saturn, and if all went well, to Uranus. (Neptune was not officially added to *Voyager 2*'s itinerary until after *Voyager 1* successfully sped past Saturn's Moon Titan in 1980.) So despite the caution of NASA planners, the Voyager project was fully capable at launch—as long as adequate programming was sent to either craft while in flight—to complete a grand tour of the solar system's four outermost planets after all. Solutions to many problems, like communication with the crafts, would have to be worked out later if Voyager's programs were to succeed over many years.

A Journey of a Thousand Miles . . .

Saturday, August 20, 1977, was a mild late-summer day. On that clear evening I was observing the sky with friends at Jarnac, our country retreat in Quebec. The Shoemakers were in the midst of a busy summer filled with Grand Canyon adventures and science. At the Kennedy Space Center in Florida, a mighty Titan Centaur rocket shuddered to life, shed its hold-down clamps, and gracefully soared into the blue Florida sky. *Voyager 2* was on its way to the outer part of the solar system.

Two weeks later an identical ship was launched as *Voyager 1*. On a slightly different trajectory, this craft would reach Jupiter a few months before its older twin. During 1978, both craft sped outward, through the asteroid belt, losing speed as they approached their first target, the giant planet Jupiter. One craft was having problems with its computer, which shut it down repeatedly. Since reprogramming the computers was possible en route, both craft were healthy by the time they reached Jupiter. The project took advantage of the wisdom and experience of several teams of scientists, but no team received more attention than the imaging science team, led by Brad Smith of the University of Arizona, and including Harold Masursky and Lawrence Soderblom of the U.S. Geological Survey.

After his stinging rebuke of NASA's Apollo program in 1969, Gene figured that his days working with space missions were quite numbered. So it was with great surprise and pleasure that he heard the words of his former student and USGS colleague Lawrence Soderblom, now deputy chief of the Voyager imaging team: "Gene," Soderblom invited, "you paid your dues on the Moon. Why not come and join us on the Voyager imaging team?" It was a first-rate opportunity. At the time of Soderblom's invitation, the spacecraft were already in the asteroid belt, their cameras all calibrated and ready to go. "All the hard work was already done," Shoemaker admitted. "It was an offer I couldn't refuse." The years of mission planning were over with. The encounters themselves were what lay ahead. With the other geologists, Gene hoped that if the surfaces of some of the moons were old enough, they would preserve a story of cometary impact activity in the outer solar system just as the Moon had for the inner solar system. He also was excited about the prospects for discovery—"you only go there the first time once," he said wistfully.[1]

A SOLAR SYSTEM IN MINIATURE

As *Voyager 1* closed in on Jupiter in the spring of 1979, the imaging team wondered what the planet's cotillion of four giant moons would be like. Called the Galilean satellites after Galileo Galilei— the Italian scientist who discovered them in January 1610—Io, Europa, Ganymede, and Callisto orbit Jupiter like a solar system in miniature. Earlier observations had already shown that Ganymede, the largest moon, had an icy surface. The big question was, would there still be craters left on it to study, or would the viscous flow of the ice have erased them? Shoemaker hoped that at least one of the Galilean satellites would have a crater record; the betting was that of all the moons, rocky Io would still have its craters.

On March 5, 1979, *Voyager I* swung by Jupiter, revealing a planetary system that was marvelously complex. As the spacecraft turned to salute Jupiter's innermost moon, Io, it revealed an orange surface without any craters at all. In fact, its surface was being repaved by sulfuric eruptions of five volcanoes as Voyager swung by. There were black lakes, calderas, and at least one volcano actually spewing out sulfur as the spacecraft tore by.

Jupiter's second moon was even more of an enigma. Glassy-smooth Europa seemed covered by a smooth surface of ice. While these two moons were fascinating, and Europa did have a few impact craters, neither had the crater record stretching back over billions of years that would provide Gene his much-needed study of the prevalence of comets in the outer solar system.

But then *Voyager 1* peered at Ganymede. "Shoemaker!" called John McCauley, one of the team members and a U.S. Geological Survey colleague, "this looks like a pretty good moon for you!" Shoemaker took one look at the first Ganymede picture and agreed. Larger than Pluto and Mercury, and almost as large as Mars, Ganymede is the solar system's biggest moon. It consists of ancient, dark areas crossed by lighter expanses distinguished by long grooves representing its more recent history. Ganymede is a careful recorder of comets that once inhabited the outer solar system. "On Ganymede," Gene explained, "we have an independent source of information of the size distribution of comet nuclei." As the spacecraft flew by this rugged moon, it recorded an ancient surface scarred with the results of impacts. Ganymede is surprisingly similar to our own Moon, with lots of craters, some of the newer ones with systems of rays.[2]

Seeing so many craters was a surprise, for this far out in the solar system, asteroids are relatively rare. Comets form by far the greatest population of impacting objects; the only asteroids that might be lurking this far out are rogue asteroids that had been thrown into Jupiter-crossing orbits by repeated close encounters with the inner planets. Once an asteroid becomes Jupiter-crossing, Gene expected that it would either be ejected from the solar system altogether, like the *Voyager* spacecraft, or would collide with Jupiter or one of its moons.

As *Voyager 1* left the Jovian system, it paid a call on Callisto. If Ganymede was rich in craters, Callisto was a crater-hunter's dream. Unlike any of the other moons, Callisto shows no evidence of any repaving by volcanic flows at any time in its history. However, a theory says that through a process called viscous relaxation, large craters slowly collapsed, unable to retain their shape over billions of years.[3] Despite that, the record for craters smaller than about 150 kilometers across should be complete virtually back to the time of the moon's formation. Craters on Ganymede and Callisto, Shoemaker believes, "tell us something about comets that is hard to find out any other way." The oldest craters may represent an early episode of heavy bombardment by planetesimal objects.[4] At least one impact feature goes back to the earliest period of the moon's formation. Valhalla is an ancient basin almost six hundred kilometers in diameter. It may be the result of an impact that occurred before Callisto's surface hardened, and as a consequence, its rim resembles a series of frozen waves.

ON TO SATURN

In November 1980 *Voyager 1* sped past Saturn and Titan, to be followed the next summer by *Voyager 2*. To the great delight of the geologists, the *Voyagers* found an utterly fascinating story on the nine large moons of Saturn. (*Voyager* discovered ten smaller ones.) These bodies orbit in a complex dance that involves both their normal orbits and the effect of impacts on the group. The inner moons—Mimas, Enceladus, Tethys, and Dione—are affected by impacts in one way, while the outer moons, Rhea, Titan, Hyperion, Iapetus, and Phoebe, are affected in quite another. The outer moons probably record the population of comets, while the inner ones were bombarded by comets and by the ringlike debris of shattered moons.

The inner satellites have a larger proportion of small impacts to large ones than was true of the cometary population that struck Jupiter's moons. The large crater on Mimas reflects the low proportion, but not absence, of large impactors. This crater, named Herschel after the eighteenth-century family of celestial explorers, taught quite a lesson in how to interpret the cratering records of relatively small moons. Gene believed that only the most recent phase of Mimas's history is preserved as Herschel and the smaller craters; previous large impacts may have repeatedly broken Mimas apart, with gravity reassembling the pieces each time in a different order.

Gene marveled at the story Crater Herschel was trying to tell. If Mimas was *almost* shattered by the impact that created Herschel, could it actually have been shattered many times in the past? And after each disruptive collision, could Mimas have reaccreted as the fragments collapsed into each other from their mutual gravity? For Mimas's and Saturn's other inner moons, like Enceladus and Tethys, the cycle of shattering and reaccretion makes some sort of diabolic sense. Their major lesson is that impact events on one might affect the others. Gene and his colleagues saw that Hyperion, for example, could not reaccrete after a major impact, because this Moon does a sort of gravitational dance, or resonance, with Titan, Saturn's largest moon. Any debris from Iapetus would be scattered by Titan, some of it crash-landing all the way over to Rhea. It turns out that Hyperion is irregularly shaped, and that it might have lost a good portion of its volume in past collisions.

Iapetus was a puzzle from the instant *Voyager* took its long-range photographs of it. This moon is tidally locked with Saturn, with one hemisphere facing the direction of its orbital motion around Saturn, the other one facing away. The bright and dark parts of this moon have been known since Jean-Dominique Cassini discovered it in 1671; it is possible that black dust from the next moon out, Phoebe, might be simply coating the leading hemisphere of Iapetus.

When *Voyager 1* swung out of the plane of the solar system to visit Saturn's great moon Titan, the last of the original Voyager objectives had been met. The reason Titan was so important to Voyager scientists is that Titan possesses the solar system's only atmosphere besides Earth's that is rich in nitrogen. Titan is a laboratory experiment in the evolution of organic materials that precede the formation of life. If *Voyager 1* somehow failed to reach Titan, then *Voyager 2* would have been targeted toward that moon also. But *Voyager 1*'s November 1980 encounter was a successful one. Before *Voyager 2* encountered Saturn on August 26, 1981, four years after its launch, its trajectory was quietly altered so that it would then swing on to Uranus in 1986 and hence to Neptune in 1989. As long as *Voyager 2*'s systems held out, and if its balky transmitter and frozen scan platform did not imperil the mission, the Grand Tour would be back in business!

URANUS

On January 24, 1986, coincidentally just two days before the *Challenger* explosion, *Voyager 2* flew past Uranus. The nature of the encounter was different from that with Saturn in that the craft could not fly near several moons. Its path had to take into account the need to get Uranus's gravity assist to push it onward to an encounter with Neptune, and so the only moon that could be studied closely was Miranda, the smallest of Uranus's major satellites. The craft did get full face-on views of Titania, Uranus's largest moon, and Ariel; both moons showed deep canyons and valleys. But the big surprise was reserved for small Miranda. Named, along with most of the other Uranian moons, for characters from Shakespeare, Miranda is, as one scientist wrote, "a bizarre amalgam of terrains unique in the solar system."[5]

Miranda looks like a giant jigsaw puzzle put together haphazardly, its pieces fitting but not really matching. As Gene and Larry Soderblom gazed at the first images, they were awestruck. Parts of Miranda resemble the ancient highlands on our own Moon, but three other parts feature oval-shaped dark regions called coronae, features not seen anywhere else in the solar system. This icy world looked like it had been torn apart by collision and then reaccreted gravitationally, and this possible explanation gave Gene a possible example of what could have happened back at Mimas. However, if one current theory is right, stresses within the Moon caused faulting and uplifting to occur on massive scales.

This brave new world may have agreed with its namesake's famous quote from *The Tempest*. *Voyager*'s first images caused the scientists to wonder, indeed, how beauteous nature is:

> O wonder!
> How many goodly creatures are there here!
> How beauteous mankind is! O brave new world
> That has such people in't![6]

NEPTUNE AND TRITON

August 25, 1989: Twelve years to the week after the start of its epic voyage, and one day shy of eight years since its encounter with Saturn, *Voyager* met its last target, the large blue world named Neptune. The individual moons that *Voyager* studied at Neptune— Triton and Nereid, and six new ones *Voyager* discovered that orbit close to Neptune—speak of something most unusual. If Neptune's system were anything like that of Jupiter, Saturn, or Uranus, then there should have been several more medium-sized, icy moons orbiting between newly found Proteus and giant Triton. However, no small moons orbit in that space, and Neptune's story speaks of the chaos that represented the early solar system.

Triton is a large body some 2,700 kilometers in diameter, just 400 kilometers larger than Pluto. Triton and Pluto were probably, at an early time, the solar system's farthest planets. When Triton, by some mysterious process, fell into orbit around Neptune, its effects on other moons would have been catastrophic. What outer moons there were probably were wrecked by collision with Triton, or more likely, were expelled from the Neptunian system altogether. One of those early moons, now called Nereid, may have moved to a distant point from Neptune, where it still orbits the blue planet. The missing moons collided with one another, reaccreting into newer moons like Proteus after Triton's orbit. Over long eons of time Proteus settled into its almost circular present orbit.

By the time *Voyager 2* left Neptune and its dysfunctional family in August 1989, it unlocked a door to our solar system's past that had not been opened before. To learn about comets, we have to observe comets. But comets in the outer solar system are too faint to follow. Thanks to the appearance of craters on these moons, we now have a record of what the comet population was like out there. But we have much more than that. If the Saturn system of moons demonstrates how cometary impacts on one moon can affect others, and if Neptune's system shows how the interjection of a large foreign body can have catastrophic effects on the moons already circling a planet, then *Voyager*'s tour showed that impacts have changed the nature of the solar system in several ways.

Voyager also left a mystery. Back at Jupiter's moon Callisto, the spacecraft photographed thirteen long chains, or catenae, of craters. The chains vary in length from 200 to 650 kilometers. Gipul Catena, one such chain, consists of about twenty equal-sized impact sites that total 650 kilometers in length. At the time of *Voyager*'s discovery, planetary geologist Quinn Passey and Gene speculated that these craters were the result of the debris sent off by a larger impact somewhere else on Callisto, causing these "secondary craters" to form. Ganymede, with a somewhat younger surface, has three such chains. These catenae were enigmas that would be put away in a file and left to gather dust. At least they would until a subsequent event in Gene's life would open the file and allow the puzzle to soar into the front pages of our study of the solar system.

CHAPTER 17

Comets and Carolyn: 1980–1995

The front of heaven was full of fiery shapes.

—SHAKESPEARE, Henry IV, *1596*

WHEN *Voyager 2* passed Saturn in August 1981, Carolyn was just beginning to get used to the long nights at Palomar Observatory as part of her husband's search for asteroids and comets. In January 1986, as the intrepid spacecraft was leaving its encounter with Uranus, not only was Carolyn fully involved with the program, but she had also discovered the seven comets listed in table 2, a remarkable achievement. (UT date in tables 2, 3, and 4 refers to the date (in Universal Time that discovery plates were taken.)

By the time *Voyager* had completed its work at Neptune in 1989, Carolyn Shoemaker had paled her earlier accomplishment. In August 1989 her total of new comet discoveries was seventeen.

Voyager's encounter with Neptune took place a few days before one of the Shoemakers' monthly observing periods at Palomar. A few days before the encounter, Carolyn asked me if I would like to observe with her at the eighteen-inch telescope. With great excitement for my first observing session with the Shoemaker program at Palomar, I prepared to leave for California on the night of August 25, 1989. On that evening I did three things. I packed for California, watched live images of Neptune's moon Triton as they flickered onto TV screens, and conducted my own visual search for comets with Miranda, my backyard telescope. Across the United States, PBS was bringing Triton images directly from *Voyager 2* into the homes of millions. I felt as though I was standing right next to the imaging team of Brad Smith and Gene Shoemaker. Outside my home that evening was a dark and clear sky, and a waiting

TABLE 2
Carolyn's Achievements, Part 1

UT Date	Comet Name	Original Designation
1 September 1983	Shoemaker	1983p
27 May 1984	Shoemaker	1984f
27 September 1984	P/Shoemaker 1	1984q
23 October 1984	Shoemaker	1984r
25 October 1984	Shoemaker	1984s
21 November 1984	P/Shoemaker 2	1984u
10 January 1986	P/Shoemaker 3	1986a

telescope. So in between views of Triton's crater-scarred surface, I went outside to search for comets like those that, eons ago, had crashed into that distant moon. And just after eleven o'clock that productive evening, while sweeping the sky visually with my telescope, I discovered my fifth comet. As it now speeds back beyond the orbit of Neptune, the comet is called Comet Okazaki-Levy-Rudenko.

It was with this new discovery, made visually through a relatively small backyard telescope, that I entered the photographic world of the Shoemaker search program. By the time I joined the Palomar Asteroid and Comet Survey, Carolyn had made a remarkable shift in her life. Ten years earlier she had been a homemaker whose family had grown up, and she had been searching for a career. Now she was the world's leading living discoverer of comets. What had happened since the end of 1982, when PACS had achieved limited success with just a handful of Earth-approaching asteroids?

FROM BARNARD TO SHOEMAKER

Since Edward Emerson Barnard found Periodic Comet Barnard 3 as a trailed smudge on a photographic plate in 1892, many comets have been discovered on photographic emulsions. Ninety years later, Gene and Carolyn were continuing in that tradition. Essen-

TABLE 3
Carolyn's Achievements, Part 2

UT Date	Comet Name	Original Designation
4 March 1986	Shoemaker	1986b
25 April 1987	Shoemaker	1987o
18 October 1987	128P/Shoemaker-Holt 1	1987z
24 November 1987	Jensen-Shoemaker	$1987g_1$
23 January 1988	Shoemaker	1988b
13 May 1988	Shoemaker-Holt	1988g
11 June 1988	Shoemaker-Holt-Rodriguez	1988h
9 March 1989	121P/Shoemaker-Holt 2	1989j
11 January 1989	Shoemaker	1989f
13 January 1989	Shoemaker	1989e

tially, their program was soaring. When the scientific collaboration of the Shoemakers and Glo Helin ended in 1982, Carolyn and Gene prepared to use the eighteen-inch telescope themselves as part of their new survey. Carolyn was uneasy about the telescope at first. "I was tremendously nervous the first time I used even a simple calculator," Carolyn remembers. "I thought I would make a mistake and the whole thing would blow up!"

Before they could hunt for anything, Gene and Carolyn had to adjust to the demanding schedule of all-night observing. Within a few months both Shoemakers were getting used to the eighteen-inch and the schedule of the night. Developing a proficiency with a telescope was quite a learning experience, one that went hand in hand with becoming comfortable with the stereomicroscope. At first Gene and Carolyn "discovered" every possible speck of dust and emulsion defect. But with time and practice, they learned to differentiate between what was real and what was not. "With experience," she says, "your eye and your mind will tell you when you have something."[1] Occasionally that something might be in an area she had looked at a few seconds earlier. Moving slowly through the field, she would suddenly realize that something was there, and then she'd move back to check.

"Carolyn went gangbusters on the stereomicroscope," Gene raved, sounding even more enthusiastic than usual. "It was a natural thing for Carolyn, who has a high natural ability to pick small things out, to recognize small anomalies." Once they got the stereomicroscope, Gene recalls, "we stopped taking long exposures, got rid of the filters, dropped down to four-minute exposures, and really started to cover a lot of sky." A lot of sky indeed: on one night they raced through the two-story observatory building, rushing up and down stairs, loading and unloading film, pointing the telescope, and guiding the telescope as it exposed the sky, for thirteen hours. By dawn the couple had obtained a record ninety-six exposures.

The Kodak type "IIa-D" film they had been using at the start of the program was very sensitive to light, but it was coarse-grained, so that its collection of faint stars was hard to decipher through the stereomicroscope because of the film's structure. They tried different films. PACS needed a knight in shining armor at this point, and it found one in the likes of Alain Maury, a French observer and photographic specialist at Palomar. In 1983, after Maury described a new film to them, Gene and Carolyn switched to Kodak's 2415 technical pan, a very slow film that under normal circumstances would require very long exposures to pick up the faint asteroids the program was searching for. Maury taught Gene and Carolyn how to speed up the sensitivity of a film by "hypering" (see chapter 1). Although exposures with the new film, even when hypered, still required six, eight, or ten minutes, the film's fine grain was capable of recording fainter objects with fewer film artifacts.

The only problem with 2415 was that its base was so thin it handled like Saran wrap. In the telescope, parts would bubble out, resulting in areas of poor focus. Soon they found the identical emulsion on a much thicker base, a film that would be less likely to buckle in the telescope. Called Kodak 4415 Technical Pan, this film became the standard for the program. Each six-inch circular piece had to be cut from an 8- by ten-inch sheet of film. Gene cut the film in one of its corners, leaving the remainders for possible later projects.

One afternoon, while scanning two films they had taken a few nights earlier, Carolyn came across two small "hyphens" in the sea of surrounding pointlike stars. Looking first at one film and then another, she realized that something was moving across those films, and fast. The asteroid was racing so quickly that even in a short exposure, it left a trailed image resembling a hyphen. Later named Nefertiti, this was an "Amor"-type asteroid whose orbit crosses that of Mars. It was the first of a long series of near-Earth objects that Carolyn would discover through the stereomicroscope. Later in 1983 Carolyn found a second Amor, designated 1983 RB and seen again ten years later.

COMETS AND CATASTROPHES

With Gene's increasing interest in comets, the couple were aware that sooner or later they should chance upon a comet. In fact, the name for the new survey program, "Palomar Asteroid and Comet Survey," was partly a response to his belief that the biggest impacts that took place on Earth, including the K/T impact 65 million years ago, were far more likely to have been caused by comets than asteroids. (The original program, we recall, was titled Planet-Crossing Asteroid Survey.) Of all the asteroids presently in orbits that cross the Earth, only one is big enough to make a crater approaching the size of the Chicxulub crater, the result of the K/T event. Its name is Ivar, but even with its nine-kilometer-wide bulk the crater it would carve out would be quite a bit smaller than Chicxulub. It is likely that 65 million years ago, while the individual asteroids that were in Earth-crossing orbits were different, their size distribution was about the same as it is now.

Ten-kilometer-wide comets make close calls to Earth, coming within a few million kilometers of it, every two hundred years on average, and their impact velocities would be considerably higher than those of asteroids, meaning that their explosive energy would be much higher ($KE = \frac{1}{2}mv^2$) than the same-sized asteroid.[2] As we have seen in chapter 6, in 1983 Gene met with other scientists involved in impact theory for an intense workshop during which the evidence for comets versus asteroids was bandied about. Gene was impressed with the amount of iridium that the Alvarez team and others were finding that blanketed the Earth from the impact 65 million years ago. So much iridium, he thought, could not have come from an object even ten kilometers across, hitting Earth at the typical velocity—thirty kilometers per second—that asteroids have in our corner of the solar system. It would take a comet that size, traveling at twice that speed, or a larger-size asteroid, to pack the necessary punch.

We really do not know the nature of the object that collided with Earth 65 million years ago. Some scientists dispute Gene's statistics, claiming instead that the numbers of large asteroids certainly do not preclude the idea that a large asteroid could have hit the Earth. Also, a large fraction of a comet's mass is water ice, which contains no iridium. If it was a comet, it must have been a very large one.[3]

FINDING COMETS WITH THE EIGHTEEN-INCH

Given this new wisdom, Gene decided that comets should be a major target of his search. In 1982, before the couple was even active in the observing part of their project, Gene and Carolyn tried to see what a faint comet would look like in their films. They chose Comet Bowell, which had been discovered by their friend and colleague Ted Bowell two years earlier and was now retreating from its visit to our neighborhood. The comet was very interesting; as a result of its encounter with Jupiter, Bowell's comet was off on a new orbit that would take it out of the solar system forever, to travel eventually among the stars. However, it did not leave a strong photographic impression. "It showed up pretty well on one of our films," Carolyn noted, "but not on the other. I thought that if they are that diffuse and faint, I'm never going to discover a comet." No new comets had appeared by the middle of 1983, but Carolyn did detect several previously known comets, including some that were found by the infrared astronomical satellite in the course of its several-month-long survey of the sky. Learning that not all comets were as difficult to see as Comet Bowell, Carolyn wondered how excited she would feel if she ever saw a comet that was still unreported.

On a warm September afternoon in 1983, Carolyn was using the stereomicroscope in the Shoemaker office at the USGS, its usual home when it was not with them at Palomar. (The Shoemaker survey is one of the more not obviously geological tasks to which the U.S. Geological Survey's Flagstaff office has been put.) The last observing period, or run, was productive; the Shoemakers took several hundred films. There were so many that Carolyn barely had time to scan any of them before the observing run ended and she and her weary husband returned to Flagstaff. As Carolyn remembered this dramatic moment—

> Gene was out of town, and I was trying to scan through what seemed like a monumental number of films before we went observing again. And all of a sudden, there it was, and I knew it was a comet. I had a feeling that it might be a new comet, but I wasn't sure. We didn't have a catalogue of comet positions at the time, or anything else to tell us if a comet was known or not. So, I hastily wrote down the approximate positions, plotted them on my star map, telephoned Brian Marsden, and said, "I have a comet."

Longtime director of the International Astronomical Union's Central Bureau for Astronomical Telegrams, Brian Marsden is at the helm of an organization that decides who gets credit for finding anything astronomical that moves, like a comet or asteroid; or explodes, like a nova or supernova. When Carolyn telephoned Brian that afternoon, he knew immediately from the positions of the comet she was reporting that it was a new one. Accordingly, the director "walked her through" the process of reporting it, including the comet's discovery positions. There was one final question at the end: "What is the magnitude, Carolyn?"

Stunned, Carolyn stared blankly. She had no experience in the complex art of estimating the brightness of a comet. "Brian," she replied, "I have no idea!" The IAU *Circulars* add the magnitude as a guide so that people wishing to observe the comet will have an idea of what its brightness is. There are accurate methods of estimating brightnesses for later scientific study, but on a discovery announcement the magnitudes are rarely determined with such care. Understanding Carolyn's sense of panic, Brian said, "It's okay! Say anything!"

Thus, the discovery of Comet Shoemaker was announced to the world, its magnitude listed (it was a good guess) as sixteen. The comet was moving in a leisurely parabola round the Sun, and nowhere near the Earth. Surely, this first comet was not in danger of striking any planet. Gene and Carolyn probably noted the extreme likelihood that neither this nor any other comet they might find would ever collide with another world.

As we have already seen, this first find was but one of a procession of comets. By May of 1986, thousands of films had already produced a plethora of interesting asteroids and comets, and the program was finally at full stride. Gene and Carolyn took some time to do crater-exploration fieldwork in Australia, and when they returned they learned that the telescope was no longer working and that their September observing run was canceled. "We were utterly appalled." Carolyn described their feeling of shock on hearing that their primary working tool had been taken out from under them. After waiting through several canceled observing runs, the Shoemakers received a frightening phone call from Alain Maury. He had heard that Palomar was about to shut down the eighteen-inch telescope permanently. Gene immediately called the associate director of the observatory and offered to pay for all repairs of the telescope and to open a fund for telescope maintenance.

The telescope was saved, but the close call left Gene and Carolyn skittish about its future. In March 1988 Gene and Carolyn and I talked about the possibility of using a second telescope, the sixteen-inch Schmidt on Mt. Bigelow north of Tucson as a backup facility if the ailing eighteen-inch were to give problems again. It was, in fact, our first meeting together. That December we completed a successful two-night run atop frigid Mt. Bigelow, northeast of Tucson. We found that the telescope would indeed work as a backup, but it was never needed for our program. (Years later Steve Larson, a comet scientist at the University of Arizona, began the successful Catalina Sky Survey, an asteroid and comet search with this telescope, using a CCD electronic detector.)

COMET TAILS

In 1987, Carolyn's eighth comet find surpassed the record of the eighteenth-century astronomer Caroline Herschel as the woman who had found the most comets. "Passing Herschel's record was a special goal for me," she said, "not because there was anything personal at all there, but because it was a landmark and special in a way to find more than any other woman had found so far." Along the road to passing Herschel's record, Carolyn passed some pretty other interesting milestones. The afternoon of September 1984, for one example, was a busy one at the Shoemaker office at the USGS. Carolyn had been scanning films all morning and then had gone for lunch with Gene. "After lunch, I went back and put a pair of films on the stereomicroscope. I looked through the eyepiece, just to see if the films were lined up, and there was this glorious comet sitting there. My heart leaped; it was faint, but to me this comet looked absolutely spectacular." Thus Comet Shoemaker 1984o was announced to the world.

Early in 1988 Henry Holt, an old friend of Gene's from Surveyor's Moon-exploration days, walked into Gene's office. "What are you working on?" Henry inquired. Gene understood how much of a toll the observing program was taking on just two people, and in short order Henry joined the observing team for occasional runs. What really excited the Shoemakers about the Holt collaboration was that Holt could handle the summer schedule while the Shoemakers were occupied with yet another new program that took them to the Australian Outback to study its well-preserved impact craters. Late in 1987 the first Shoemaker-Holt comet was announced. On June 24, 1988, Carolyn's fifty-ninth birthday, Carolyn began scanning a pair of films that Henry, his son Hank, and

Tim Rodriguez took at Palomar in the Shoemakers' absence. Anxious to see if Henry's films had captured an asteroid she was hoping to follow, Carolyn asked if she could examine his films. In no great hurry to scan a field of sky that was dense with stars in the richest part of the Milky Way, Henry happily complied. When films are crowded with stars, scanning is difficult and takes much more time. "The films were almost wall-to-wall stars," Carolyn explained. "I was plowing along, hoping to find that asteroid." In less than an hour she found the faint speck of light, ever so slightly shifted from one film to the other, from the asteroid she sought. She plotted its position and then kept on looking. A few minutes later she found the telltale fuzzy images of a comet. This was a new comet, it turned out, and soon they were on the telephone with Brian Marsden to announce Comet Shoemaker-Holt-Rodriguez. That was the year, it turned out, that Carolyn had an unexpected comet guest at her birthday dinner.

Hyphens in the Sky

Of Gene's multitude of varied projects, the observing effort was special for a number of reasons, the most important of which was that Carolyn was a part of it. Observing is fieldwork, even though the samples are a long way off and can be collected only as their light is captured on film. But fieldwork is what the couple remembered most fondly about the early years of their marriage, and by 1983 it was quickly apparent that the couple had found a way to relive their early time together.

In November 1989, as we have seen in the Preface, I had a chance to see this closeness personally. I was guiding an exposure at the telescope when Carolyn opened the door to the dome and called out "Got a fast mover!" Gene rushed downstairs to look at the two hyphen-shaped images on the discovery films we had taken the night before. Things got very busy at that point. We had to continue the night's program of more than fifty exposures of specific fields of sky. As the Earth was turning, star fields were moving out of optimum places, so we could not pause—at a time like this, I remembered my father's words, "the stars wait for no man." Once the new field was in the program, things calmed down as the night's work continued. Although it was true that the stars didn't wait for us, they seemed to shine brighter than usual that night. As I was guiding the telescope, Gene walked up with a new film for the next exposure, and as he passed Carolyn, he hugged her.

We also had to follow up on the new discovery at once. While Carolyn and I kept feeding films into a telescope hungry for images of the sky, Gene added the previous night's discovery field to this night's program. And then as Gene guided the telescope on another exposure, I used my laptop computer to find the positions of an appropriate guide star (a star to track the telescope on during the exposure) near the new asteroid.

At the end of the exhausting but happy night, with two nights' observations of the asteroid, we telephoned it to Brian Marsden. A few days later on the basis of more observations, he determined that the new object was a member of the rare Aten group, with orbital periods less than a year.

If You Don't Look . . .

On one winter night the cold wind was blowing so hard that we tried to keep the telescope pointed east, and out of the wind. As the night progressed, our search program called for a field far to the northwest, so we had to brave the gale. Although I tried to keep the telescope steady, the thin metal shutters at the top caused it to shake like a sail. The guide star was dancing about in the eyepiece field, and I was unable to center it. Then Gene had an idea. For the next exposure, he stood precariously on top of an elevating chair, reached to the top of the telescope, grabbed the shutters, and held on tight. Perhaps, we hoped, Gene's balance and strength would hold the shutters, and hence the telescope, steady enough. I swung the telescope into place. As Gene's chair rose several feet, he stood on it to reach the top of the telescope. I peered through the eyepiece at the guide star, which was dancing about like a firefly. All was

TABLE 4
Carolyn's Achievements, Part 3

UT Date	Comet Name	Original Designation
17 September 1990	P/Shoemaker-Levy 2	1990p
15 November 1990	P/Shoemaker-Levy 1	1990o
22 January 1991	Shoemaker-Levy	1991d
7 February 1991	129P/Shoemaker-Levy 3	1991e
9 February 1991	118P/Shoemaker-Levy 4	1991f
2 October 1991	P/Shoemaker-Levy 5	1991z
6 October 1991	Shoemaker-Levy	$1991a_1$
7 November 1991	P/Shoemaker-Levy 6	$1991b_1$
13 November 1991	P/Shoemaker-Levy 7	$1991d_1$
5 April 1992	P/Shoemaker-Levy 8	1992f
25 October 1992	Shoemaker	1992y
24 March 1993	P/Shoemaker-Levy 9	1993e
23 May 1993	Shoemaker-Levy	1993h
14 March 1994	Shoemaker-Levy	1994d
14 May 1994	P/Shoemaker 4	1994k

ready: I pulled the lever and the shutters slowly opened. As Gene grabbed on to them, the shivering guide star settled down. It was still moving around, but not as much. For the next few minutes the wind howled as Gene held on to the shutters with his hands and to the chair with his feet, trying to keep the scope steady and himself from tumbling off the chair. Meanwhile I tried gamely to keep the guide star centered. Scanning films downstairs, Carolyn heard the commotion and climbed the stars. "Is this exposure absolutely necessary?" Carolyn asked.

The noise from the wind was so loud that we could hardly hear our own words. "*Is this exposure absolutely necessary*?!" she repeated.

"*Yes!*" Gene hollered back, his voice still difficult to hear through the pounding wind. We imagined his thought after that: You never know what's in the field if you don't shoot it.

About forty-five minutes later we had to repeat the acrobatic exposure, but then we turned the dome to the east and continued observing in the more sheltered areas. As Carolyn began scanning the developed films, she noticed a faint, fuzzy object moving across the sky. Crossing the very field we had gone to such an extent to photograph was a new comet that became known as Shoemaker-Levy 7.

CHAPTER 18

A Rock-Knocking Geologist: 1984–1995

Why then the world's mine oyster,

Which I with sword will open.

—SHAKESPEARE, Merry Wives of Windsor, *circa 1597*

HAD CAROLYN not wanted to observe with Gene, her husband said many times, their Palomar Asteroid and Comet Survey would never have happened. As the observing program was gaining speed in 1982 and 1983, Gene realized that he was back where he wanted to be—doing fieldwork, or at least a sort of fieldwork. Already in his middle fifties, he understood that if he was ever to return to the "rock-knocking" fieldwork that he had loved in his youth, he would have to plan it soon, while he and Carolyn were still at the peak of their strength. And so germinated an idea, one that would beautifully connect two major strands of his research. On one side there was the future: a survey of comets and asteroids that could strike the Earth. On the other was the distant past: going back into time to study the effects of impacts on Earth in ancient years.

AUSTRALIAN OUTBACK: A HOLY GRAIL
OF IMPACT SITES

Ever since Gene first peered into Meteor Crater in 1952, digging into old impact sites had been a top priority. Thanks to the ever-present processes of erosion and weathering, Earth's record is poorly preserved, especially from sites dating back into Proterozoic time. To discover impact sites dating back so long, a geologist needs a continent known for long-term stability. Gene knew there was

25. Gene at Wolfe Creek, 1990.

one such continent, Australia. It is smaller than the other major
landmasses and separated from them by vast expanses of ocean.
Antarctica's ancient impact structures were quite beyond reach,
buried as they were under kilometers of ice. Australia, a pleasant
land with dry deserts, provided the golden opportunity Gene
sought. With little orogenic, or mountain-building, activity in
billions of years, it should boast, Gene thought, a host of well-
preserved ancient records. Gene also noted that since most of the
Australian geologists were employed in the search for mineral resources, not impact sites, here was a place where he could work
without interfering with local programs.

The Australian idea began in the late 1970s, with a picture someone sent to Gene of Wolfe Creek Crater. At some 300,000 years of
age, it is one of the youngest and best preserved of Australia's impact sites. Although it is about the size of Arizona's Meteor Crater,
it is six times older, and so is more eroded. Carolyn viewed the

crater as a picture of stunning desolation. "I would not be thrilled to see such a place," Carolyn noted in her first reaction.[1] To Gene, though, the desolation was magnificent. In 1983 he wrote a proposal to NASA for three months of annual fieldwork, the purpose of which was to determine whether the rates of impacts on Earth have been stable throughout geologic time, or whether they have changed. The proposal was approved by NASA, minus (typically for NASA) funds for data reduction and analysis, and by Carolyn, who had a change of heart and wanted to accompany Gene to Australia.

PLANNING FOR DOWN UNDER

Over the years, Gene had amassed files on many different impact sites, and he turned to them to plan his first Australian assault, set for the (northern hemisphere) summer of 1984. Having little idea of the vast distances and difficult driving conditions in the Outback, Gene optimistically planned to hit Western Australia one year, and the Northern Territory the next, so that in a few years the project's fieldwork would be done. With experience, they learned to plan each year's trek from north to south, to take advantage of the slowly warming weather of the Australian spring.

In June 1984, the couple ended a run at the eighteen-inch, returned to Flagstaff, packed their bags, and were off for a new world. After arriving in Perth, Gene and Carolyn set up shop at the headquarters of the Geological Survey of Western Australia, where they spent almost a month poring over images of Australia taken from space by the Earth resources satellite *Landsat*. Aerial photographs supplemented for closer views, but some sites were so remote that they hadn't even been surveyed by air. The Shoemakers amassed a library of topographic maps, geologic maps, and ordinary road maps, rented a vehicle, and set off into the Outback. "We wanted to build on what the Australian geologists had done before." Carolyn recalled their optimistic note as they set off.

That first season was a real lesson from the school of hard knocks. The Survey in Western Australia had outfitted their vehicle with a radio phone, with which Gene reported slower than ex-

26. The Toyota Hilux near Connolly Basin, an impact structure less than 60 million years old.

pected progress toward their goals. Although the radio was a useful tool, even *it* caused problems when the long and expensive aerial fell off. This mishap cost the couple two days of thorough searching with Gene driving slowly while Carolyn sat on the hood of the car, looking for a glint of metal, which she finally found in high grass. The sites they saw—Spider, Picaninny, Connolly Basin, Veevers, and Teague Ring—had to be considered just preliminary scouting.[2] In other years they would see other sites, like Henbury, a pattern of several craters—the main one is a double, and it is surrounded by nine others. The impacts that produced the complex happened just a few tens of thousands of years ago.

When the Shoemakers planned their trip, they realized that they would need to be well equipped to spend weeks at a time in remote areas. On Friday, July 6, 1984, Gene and Carolyn stopped at their first structure, a site called the Dampier Archipelago. Suggested to be possibly impact in origin, it beckoned to the couple. Carolyn wrote in her journal, "We camped this evening on a sand dune at Hearson Cove, a beautiful and romantic setting with red blooming flowers arranged around our site and the sounds of the Indian Ocean lapping on the nearby shore. Above us," she concluded prophetically, "faithful Jupiter twinkled down and the Moon lit our beach." Accessible only by sea, the island site, which the Shoemakers shared with sea turtles laying eggs, was magnificent. But Gene's research confirmed the sense of local geologists: this was not an impact site.

Moving onward on their trek, they encountered Australia's most famous animal: "We saw our first wild kangaroo hopping across a hillside," Carolyn wrote. "We also had our first flat tire, a new tire with a stick right through the tread!"[3]

A Vehicle for All Seasons

By 1986 the couple had decided to move their equipment down to Australia more-or-less permanently, to buy rather than rent a vehicle, and to design its space so that it would hold their lives while they were there. The Toyota Hilux was a small, reliable four-wheel-drive pickup vehicle designed for rough travel, which they outfitted for tumultuous travel. With three sides in the back that folded down, the truck was easy to pack and unpack. In two specially designed plywood boxes the couple stored their belongings and tools, including an alidade and tripod for mapping, a mapping board, a gravimeter for detecting the gravitational anomalies or variations that characterize an impact site, a magnetometer for detecting iron meteorites, a theodolite for obtaining positions relative to the stars, and a long aerial for their radio. Outside of the boxes were two fifty-gallon fuel drums, and ten five-gallon containers of water. Since much of their driving was off road in remote areas, they also carried a large "Kangaroo jack" and a long length of chain—equipment strong enough to pull a large truck out of a ditch—and a stake to plant should no trees be around to help pull the vehicle out. They also had a smaller "come-along" winch to rescue them from less serious problems. Hanging off the side of the truck were eight rubber doormats that helped in navigating over sand dunes and creek beds. In case the battery died, robbing the couple of their transportation *and* electrical system, the vehicle had an extra battery.

Finally, there were two spare tires that along with a collection of patches and inner tubes prepared them for virtually anything, including the year they had twenty-three flats. "Gene became a real expert in repairing tires," Carolyn laughs. Sometimes a flat tire was virtually impossible to pry from its rim. To remove the tire, Gene laid it flat on the ground. Then he and Carolyn stood atop the tire, held hands, and danced vigorously on it! The tire disco eventually freed the tire from its rim.[4] At the end of a field season, they would usually arrive at the storage area with four patched tires and no spare.

There was a time that a road was so bumpy that the engine loosened from its mount. Gene rebolted it temporarily, and later he welded it to the mount so that future shocks would not unhinge it. Another afternoon, the Hilux's drive train fell off, its bolts utterly sheared away. They set up camp right where they were. Gene could not find any bolts the right size, and it looked as though they were in trouble until he cleverly "fattened" the bolts he did have with bailing wire. The arrangement saw them safely through hundreds of kilometers, all the way to a shop with the proper bolts.[5]

As it was at Palomar, a cassette player was a must, especially one that could play music that could be heard over the roar of the engine and the wind. Their music emphasized Australian singers Slim Dusty, whose rustic songs painted interesting stories, and Ted Egan, whose familiarity with Aborigine folk lore often found its way into his songs.

Living Remotely with Style

"We needed to stay in Australia long enough to do some good," Carolyn reasoned. "We felt that if you want to do something badly enough, you just carve out the time to do it." Knowing how busy the Shoemakers were, much of the geology and planetary science communities were surprised that they managed to find the time to invest in Australia. "It was especially important that we not bring along a laptop computer so we could keep up with our e-mail," Carolyn insisted. "We wanted to separate ourselves from the demands back home." The philosophy didn't always work—over a three-month period, proposals had to be written and papers were due. The fact that the world did not stop turning had two major effects on the traveling couple. First, the week before departure for Australia was hard on Carolyn. Trying to dot the i's on half a dozen pieces of writing in just a few days while at the same time packing for a transoceanic three-month trip, was a trying experience. "I don't look forward to the day before we leave for Australia," she frequently said. "But once we're on the plane, we look ahead to three months away from all that frustration."

Once down under, if some paper or proposal were due, they would write either in a motel room or right in the field. For half a day they would set up in some protected spot, open their portable typewriter, and forge ahead. At the next post office, the completed work would then be sent the fastest way possible to Caltech, or in later years to the U.S. Geological Survey in Flagstaff, where others would copy the material into the proper form and submit it, hopefully before a deadline. Even in the most remote parts, Gene wanted to enjoy his day. On one evening a passing driver, seeing the couple's truck stopped, stopped to make sure they were all right. He found the couple dining, sitting on stools at a table by candlelight, with plates, cutlery, and glasses filled with wine. In tune with Nature and with their souls, husband and wife were coaxing and cajoling the Earth to reveal its ancient secrets.

The Australian experience continued annually until 1997. The couple repaired perhaps two hundred flats, pulled themselves out of rivers, mud holes, and deep sand. Frequently Gene would catch illnesses of one sort or another. The only years that the Australia work was canceled were 1988 and 1994, the first to catch up on other work, and the second due to the demands of Comet Shoemaker-Levy 9. In 1996 they made two trips—one in February on the southern coast, to search for and map the locations of tektites, and the other their annual field excursion.

Windy Corner: A "Typical" Few Days in the Outback

Any field experience in a strange land is bound to have a plethora of unusual happenings. Saturday, August 18, 1984, the Shoemaker's thirty-third wedding anniversary, is a good example of a day rich with experiences in uncovering the nature of a Western Australia site that had been listed as possibly related to impact. The day was typical in that it was busy, but it began with uncharacteristically cloudy weather. "Off to an early start," Carolyn wrote in her journal that day, "but then a flat tire! Called for a repair job since we have no good spare at present. The country looks like Colorado Plateau, red sandstone, trees shaped a lot like junipers and widely spaced, shrubs with the coloring of sage brush." They reached a town called Meekathera, the only place within hundreds of square kilometers large enough to stock serious supplies for this excursion, but the town was essentially locked up for the weekend. They could get gasoline for their car and drums, but the one garage still open did not carry new tires for the Hilux. They were lucky enough to find one new spare tube and some tire patches. They left the town cautiously, hoping their fully loaded vehicle would avoid further flat tires.

Quiet as it was, Meekathera served as their "jumping off point" for the 1,000-kilometer drive to Windy Corner, near one of the structures that was listed as a suspected impact site. In the United States, such a drive along a big interstate highway shouldn't take more than nine hours at the wheel. And even here in the Outback, the first part of their wedding anniversary drive to the village of Wiluna, was arrow straight and relatively smooth. Under gathering rain clouds, they drove as late as they could and camped for the night under a tarp.

By Monday morning, August 20, the couple had progressed closer to their goal over a road covered with sand dunes, washouts, and rocks. The storm having moved on, the couple awoke to a clear morning and completed the drive to Windy Corner. They gave the site at Windy Corner a quick inspection. It was a large, somewhat wooded area, some ten kilometers wide, and featured a barely discernible rim surrounding a circular depression. On first inspection, it did not look promising as an impact site. On one rim, they found a small outcrop. "The sandstone looked like it might have been crushed," Carolyn wrote. With no other supporting evidence, Gene gave it a 5 percent chance of being the result of impact.

"As we finished lunch and were about to get underway, we saw that we had another flat tire! Another repair job with Gene teaching me the whole process. Finished, we drove down the track which cuts across the structure." They found a convenient seismic track, cut for oil exploration, that cut across the center of the structure. "We turned and drove south, and then, joy!! a rocky ridge in the center with fractured and sheared rock. We spent the rest of the afternoon looking at it, at which point Gene raised the odds of impact. We were looking at the central peak!"

That evening Carolyn labeled and wrote brief descriptions of each of the samples that had been collected. In the morning they began a cross-country drive through the Windy Corner structure, recording as much as they could about its central peak. "Navigating by mileage, map, and air photos, we turned off on the old Wapet-Uranus seis [*sic*] line." Eventually, the site near Windy corner was confirmed as the site of an impact that occurred somewhat less than 60 million years ago. The Shoemakers named it Connolly Basin after the grandfather of a geologist they knew, a man who had helped pioneer that section of Australia.

Still having miles to go, the Shoemakers pushed on to their next target: Veevers, a site lying between the Gibson Desert and a large expanse of desolation appropriately called the Great Sandy. "This morning we took the metal detector to find meteorites," Carolyn wrote on August 21, after the couple gave the site a cursory inspection and put it high on their list to return to and map in 1985. "Once set up, he pointed me where to go, and I hunted for about four hours. Eureka! Found one abt. 10 grams and a couple of pieces highly oxidized. We decided they were really sparse and anything big may yet be embedded in the crater. Nevertheless, with our discovery we can truly call Veevers a meteorite crater. Ours were the first found there. Gene would like to come back and do a geologic

map here—would take about a day."[6] They did return in a later
year for several days of mapping. Gene later determined that the
site was an impact area slightly younger than a million years.

THE AUSTRALIAN CRATERS EXPEDITION

In 1990 the American Meteoritical Society had its annual meeting
in Perth. As part of the extended events for this meeting, Gene and
Carolyn arranged a three-week-long craters expedition across the
Australian Outback. Before the trip, the Shoemakers led a small
group backward along the planned route, from north to south,
from Alice Spring to Perth across the Western Australian desert
along the "Canning Stock Route." They were accompanied by
other geologists, including Candace Kohl and Kuni Niishizumi.
Gene was working with them to develop a way of dating impact
craters. The theory: If one could find a spot on the inside crater
wall where the rates of erosion and deposition canceled each other,
and then determine how long the spot had been exposed to cosmic
rays, one could then determine the age of the crater.

One evening, the group pulled into Boxhole crater, an impact
site estimated at some thirty thousand years old. The almost full
Moon made the sky and the desert so bright it gave Gene an idea.
"Who wants to see the crater in the moonlight?" he asked. The
small group walked around Boxhole for two dark hours—a crater
on Earth lit that night by the crater-scarred full Moon. "Walking
around that crater in the moonlight was just marvelous," says
Kohl.[7] Gene had a way of making the pursuit of science a very
romantic process.

Once the main expedition began, ferrying fifty scientists through
the Outback had major problems. Consisting of a twenty-seven-
passenger bus specially designed to climb dunes, five Land Rovers,
and a supply truck, the caravan made its way from Perth toward
Adelaide. It turned out that the only way the bus could climb the
dunes it had been designed for, was inch by inch. The bus took so
long to crawl its way up one dune that Gene realized it could not
possibly traverse 350 dunes in this way. The result: a detour of

27. The Meteoritical Society Tour in Australia, 1990. Gene at Dalga-ranga, in Western Australia. The impact site is some 27,000 years old, but the beard is new.

more than fifteen hundred kilometers. Participants were placing bets on whether Gene could complete the tour in time for them to catch their flights from Adelaide to Perth. That evening the windshield vibrated out of the supply truck and ended up sitting on the laps of those sitting in the front seat. The group camped at Well 33 that evening, and then began its long detour to the coast the next day.

Even with soured moods, Gene had a way of persuading the group to take the events in stride, thinking of them as part of this marvelous adventure of discovery. Despite the problems, the trip seemed a great success. According to one participant, this was Gene's effective and informal way of disseminating the results of fieldwork to a broad audience. After six years in the Outback, Gene had succeeded in presenting the Shoemaker findings to many geologists, all without having to write a peer-reviewed paper!

Of Craters and Radiator Faxes

By the end of the Shoemakers' Australian odyssey, Gene and Carolyn had succeeded in studying a plethora of craters. The Northern Territory's Kelly West is a site that shows a few remains of an impact far back in the Precambrian. Gosses Bluff, another Northern Territory impact site, takes up more than thirty kilometers, and has been precisely dated at 142.5 million years.

In 1996 David Taylor of York Films, while filming for his documentary *Three Minutes to Impact*, arranged to get Gene and Carolyn a satellite phone that was used to make arrangements for their visit to a billion-year-old site called Liverpool, some three hundred miles from the city of Darwin. "This is a crater I've been hoping to get to for a dozen years," Gene said excitedly in 1996. Since the site was in Aborigine territory, the couple, accompanied by a York Films camera crew, needed a special permit, a helicopter, and an Aboriginal guide. Shaped like a bowl, this crater has been eroded to well below the level Gene was familiar with at Arizona's Meteor Crater. "It is a snapshot of the geology that's below the level of exposure at the younger crater. We see what's happening deeper down."[8] They used the phone to contact, among others, astronaut Tom Jones, who was then preparing for an Earth-mapping shuttle mission. Based on the Shoemakers' remote field observations, Jones was able to suggest from another remote area—one hundred miles *above* the Earth—that the large area was a barely noticeable site that could be an impact crater. That year the couple also invested in a global positioning system to help them navigate a particularly difficult three-day excursion across sand dunes, and no road or track, to reach a potential impact site suggested to them by the Australian geologist Andrew Glikson on the basis of magnetic anomalies. Even with the satellite phone, no one could call *them*, and for that the Shoemakers were thankful as they visited Western Australia, the Northern Territory, South Australia, Queensland, and New South Wales.

In August 1995 a TV production team was filming me at Palomar Observatory, and they asked if they could speak to Gene. They were skeptical when I assured them that the Shoemakers could not be contacted, except for messages that could be left via their daughter Linda, and then only once every few weeks. "It's the nineties," they argued. "Anybody can be contacted at any time."

Eventually I admitted that even though he preferred to have no communication with outsiders during his sojourn in Australia, Gene did have a secret means of communicating. He designed the system himself to run off the heat generated by the Toyota Hilux's engine. The producers looked interested, so I built the story a little. "It's called a radiator fax," I explained. "Gene attached it with rustproof bolts to the radiator mount."

"Can we reach him this way?"

"Well," I added, trying desperately to look serious, "it's difficult. Once you send the fax, a special buzzer goes off in the car's engine. If Gene hears it, he must then pull off the road, turn off the engine, open the hood, and let the radiator cool just enough so that the heat will impregnate the fax paper without setting it afire. The paper then rests in the radiator fax until the message is complete. Gene then carefully lifts the paper out of the machine, trying not to burn himself or tear the paper, and then Carolyn sets up a clothesline (there really was a clothesline) to dry the paper for half an hour before he can read it. After that, the clothesline comes down, Gene closes the hood, and they drive off."

The producers were so anxious to reach Gene that they really never did understand that I was leading them on. Weeks later, when I inquired of Gene as to the success of their radiator fax, he howled with laughter. "One night Gene woke me up, laughing and laughing," Carolyn said. "When I asked him what was so funny, he said 'David and his radiator fax!' "[9]

Memories

Not satisfied with being a near-perfect record of impacts, the lands around the Victoria coast provide evidence for a major impact that took place somewhere around Indochina. The evidence is in a mineral called australite, in the form of a glassy tektite that was thrown up from the impact site and fell back to second landings in Australia. With tektite collector Ralph Uhlherr, the Shoemakers collected samples along the high cliffs of Victoria's south coast. Despite the beauty of the coastline, the work could be treacherous, for the tektites were most abundant on steep slopes that ended as hundred-foot-high cliffs.

Whether Gene and Carolyn were looking for specimens, or studying the wide terrain of a suspected impact site, or visiting with one of the many friends they made, the Australia project engendered a very happy and relaxed feeling. Sean, the one grandchild who accompanied Gene and Carolyn to Australia one year, recalls that the trip gave him the chance to "learn how down to Earth my grandfather really was. It was exciting to know that he was a great scientist, but to me he was still my grandfather." And like his mother before him, Sean saw his grandfather as a teacher. "I'd turn over rocks to look for insects and snakes, but he'd explain the rocks to me."[10]

In Carolyn's journal record of their first visit to a Precambrian impact site called "Spider" for its resemblance to a multilegged creature, she describes this feeling well: "Gene loves geologic field

28. Mapping channels with tektites on the southern coast of Australia.

mapping and is in his element and blissfully happy doing the thing he thinks he does best. I am enjoying being with him. There are no stresses and strains for these 5 days, so it is a real period of R and R. . . .

"Gene was up at 5:30. I could hardly believe it because it was cold again, just getting light, and firewood had to be gathered. The birds were so cold that only a few had sung, but Gene was whistling away. How delightful if this could carry over to home!!"

The twelve seasons of Australian experience tied in well to the asteroid and comet search. It was closely related, Carolyn wrote, "to our 'Mom and Pop' observing program, so-called because the monetary support was very small and for a while the two of us were the only ones involved. Much of the year we looked up to find those objects that could collide with Earth in the future. Then we took about three months in Australia to look down at the Earth to see where such objects had struck."[11] One of those impact sites,

Teague Ring, is among the oldest on Earth at 1,685 million years in age. In 1997, on the recommendation of the Geological Survey of Western Australia, Teague Ring became the Shoemaker Impact Structure.

For a long while after each return from Australia, Carolyn and Gene would still be thinking down under even while trying to adjust to the strenuous pace of life in the United States. I noticed that for every first observing run after an Australian trip, I could expect to hear a major dose of the music of Ted Egan and Slim Dusty. As Carolyn wrote at the end of her 1984 journal, "For some time my dreams at night were all of Australia, and many times I wandered there and traced our route while at the telescope on Palomar."

CHAPTER 19

Springtime on Jupiter: 1993

Hung be the heavens with black, yield day to night!

Comets, importing change of times and states,

Brandish your crystal tresses in the sky . . .

—SHAKESPEARE, Henry VI, *circa 1590*

EVERYONE who knew Gene Shoemaker was amazed at how the man could divide his time among so many different projects and yet remain centered on one theme. Before the twenty-third of March 1993, it was not easy to demonstrate this unifying focus. On that partly clouded evening, we took two photographs of an area of the sky on which, two days later, we discovered Comet Shoemaker-Levy 9. The story of S-L 9's discovery is the subject of chapter 1, and it began a period that utterly dominated our lives for the next sixteen months. "Without a doubt," Carolyn told an interviewer more than two years later, "the most exciting time of my life was the discovery of Comet P/Shoemaker-Levy 9."[1] On March 26, *Circular 5507* from the International Astronomical Union, announced the beginning of this story:

COMET SHOEMAKER-LEVY (1993e)
Cometary images have been discovered by C. S. Shoemaker, E. M. Shoemaker and D. H. Levy on films obtained with the 0.46-m Schmidt telescope at Palomar. The appearance was most unusual in that the comet appeared as a dense, linear bar about 1' long and oriented roughly east-west; no central condensation was observable, but a fainter, wispy "tail" extended north of the bar and to the west. The object was confirmed two nights later in Spacewatch CCD scans by J. V. Scotti, who described the nuclear region as a long, narrow train about 47" in length and about 11" in width, aligned along p.a.

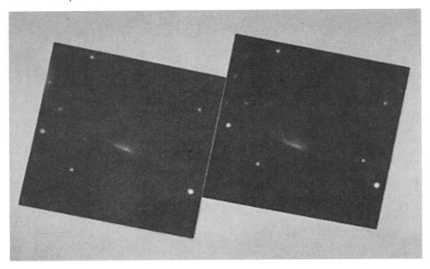

29. The two discovery photographs of Comet Shoemaker-Levy 9.
These two images were taken by Gene and Carolyn Shoemaker and
the author on March 23, 1993 (March 24 Universal Time), using the
eighteen-inch Schmidt camera at Palomar Mountain Observatory.
The picture on the right was taken almost two hours before the one
on the left.

> [position angle] 80–260 deg. At least five discernible condensations
> were visible within the train, the brightest being about 14″ from the
> southwestern end. Dust trails extended 4′.20 in p.a. 74 deg and
> 6′.89 in p.a. 260 deg, roughly aligned with the ends of the train and
> measured from the midpoint of the train. Tails extended more than
> 1′ from the nuclear train, the brightest component extending from
> the brightest condensation to 1′.34 in p.a. 286 deg. The measure-
> ments below refer to the midpoint of the bar or train. . . .
>
> The comet is located some 4 deg from Jupiter, and the motion
> suggests that it may be near Jupiter's distance.[2]

SOLVING THE RIDDLE OF CATENAE

In the first two months after discovery, surprise after surprise kept
appearing in print, or in informal hallway conversations and e-mail
notes. One of these happened a few days after our discovery, when

30. Using the 36-inch Spacewatch camera atop Kitt Peak, Jim Scotti took this remarkable image of Shoemaker-Levy 9 on March 30, 1993. Courtesy Jim Scotti and Spacewatch.

planetary scientist H. J. Melosh met Jim Scotti at the Lunar and Planetary Lab. "That picture of S-L 9 that appeared in the paper"—Mellosh was describing a photo taken by colleague Wieslaw Wisniewski—"reminds me of a chain of craters."[3] Chains indeed: *Voyager* spacecraft had imaged such chains, or catenae, on Callisto and Ganymede. Callisto, the solar system body with the most complete cratering record, has no fewer than thirteen such catenae; Ganymede has three. Other chains, like Davy Rille on our own Moon, were also geological mysteries. As we have seen in chapter 12, Gene himself thought that Davy Rille was probably the result of volcanic activity, and thus it was high on his list of must-see lunar sites.

S-L 9's earliest telescopic appearance solved the mystery of the catenae. It suggested a process during which a comet would approach within a planet's Roche limit and break apart. In its most common iteration, the process would end there, the comets continuing in their new orbits about the Sun. In another scenario, the comets could continue to circulate around Jupiter for a while before escaping into solar orbit. But at least sixteen times in the history of Jupiter, and probably hundreds more, that did not happen. Comets that broke up in this manner went on to collide, within a few days, with Callisto or Ganymede.[4] In the case of the Moon's Davy Rille, a comet could have broken up in a very close pass to the Earth, then collided with the Moon a day or two later.[5]

On April 3, IAU *Circular 5744* contained another surprise. The proximity to Jupiter made S-L 9's orbit extremely difficult to nail down, but it was likely "that the object is at least temporarily in orbit about Jupiter."[6] Extraordinary: Our comet was the first to be caught in the act of orbiting a body other than the Sun. At this time, the comet was about to be formally awarded the name Shoemaker-Levy 9 in honor of its being the ninth comet of the type that returns every few years—or the ninth short-period comet—discovered by our team.

My Darling Clementine

During this early period, it seemed reasonable that the comet would provide some initial excitement for a short while, but that as the fragments moved farther apart and faded, interest in it would decline. In the course of Gene's career, this comet seemed at first to be a sideshow, albeit a most interesting one. Gene was planning to spend much of 1993 and 1994 revisiting an old love—spaceflight to the Moon. Planned by the U.S. Department of Defense, a mission called Clementine went from drawing board to launch in little more than a year, a virtually unheard-of accomplishment. *Clem-*

entine would travel to the Moon, map it for several months, and then head off to the Earth-crossing asteroid Geographos. Gene had thought long and hard about whether he would participate, and then typically got his proposal in just before the deadline, just as he was leaving for the March observing run that produced the discovery of S-L 9. A fortnight later, he was asked to lead the spacecraft's science team.

Several of our 1993 observing runs were punctuated by long conference calls with the rest of the team, and Gene looked forward to spending a pleasant spring in 1994 in the Virginia city of Alexandria, in the closed Clementine control center nicknamed the "Batcave." At least that's how Gene's calendar looked before the end of May.

May 22, 1993: Day of Destiny

On May 20, Gene and Carolyn and I met at Palomar for our observing run. The first two nights were somewhat spotty, with clear parts being interrupted by fog and clouds. Shortly after midnight on the morning of May 22, the sky cleared, and we took photographs until dawn. On the morning of May 22, we photographed a field far to the south and east of Jupiter and S-L 9. We also continued our program, which that week would include several follow-up images of our new comet, whose fragments, moving apart at a runner's pace of about three meters per second, had stretched the length of the train so that it was now almost twice as long as it had been at discovery. We awoke later that day to return to the observatory; although we did not know it as we drove toward the little dome, Gene was about to have a meeting that would bring his life's work full circle, in an event that would touch his soul.

By dawn we obtained several dozen exposures of a large part of the sky. As was usual during our observing runs, after catching some sleep after dawn, we were up by noon and by about two o'clock we were settled in at the observatory's first-floor working area. Carolyn turned on her stereomicroscope and began scanning for new objects, while Gene moved to the solitude of the dark room. I opened my laptop to read e-mail, particularly to see if any new objects had been found for which we would need to perform follow-up observations. As the modem's scratchy whine noted the developing online connection, I could hear a loud thump every minute or so as Gene slammed the six-inch circular film cutter down on yet another sheet of eight by ten Kodak 4415.

I scanned my list of e-mail messages; there were the usual annoying jokes, trivial announcements, sale offers, and a few personal messages. But my eye stopped at two missives from the Central Bureau for Astronomical Telegrams: Circulars 5800 and 5801. Usually these messages listed a variety of subjects in their brief tables of contents; this time both listed just one: Comet 1993e. "Something's going on with our comet," I told Carolyn.

I read the first message. Brian was making a critical announcement in exactly the way he said he would—years earlier I had asked him how he would announce a possible collision of an asteroid with the Earth. His answer: he would publish an orbit and an ephemeris that showed decreasing values of Delta (the object's distance from Earth) until Delta was less than the radius of Earth.[7] And here I was, reading precisely that message, only it was Jupiter, not Earth, and it was not an asteroid but our comet he was writing about. I was aghast. Carolyn looked up from the stereomicroscope as I told her the news. "David," she replied, "we're going to lose our comet!"

Inside the darkroom, Gene stopped cutting film. "What's happening?" he demanded. I answered with the words that summed up his career: "Our comet is going to collide with Jupiter."

"When?"

"In July of 1994."

What followed was a cacophony of sounds, of slamming covers and doors; Gene shoved the film he was cutting into the light-protected ammo box we used for storage and stashed away the remaining film. Making a beeline for the door and my computer, he joined Carolyn and me at my laptop. For a minute or so we just read the announcement. In that short time its implications began to sink in. Since 1956—for almost forty years—Gene had carefully tried to understand the significance of impacts on Earth, the Moon, and other worlds in our solar system. If the orbital information on those circulars was correct, Nature was about to render Gene's work a stunning confirmation. "I don't believe it," he muttered. "In my lifetime, we are going to see an impact."

Excerpts from the two circulars follow:

> Almost 200 precise positions of this comet have now been reported, about a quarter of them during the past month, notably from CCD images by S. Nakano and by T. Kobayashi in Japan and by E. Meyer, E. Obermair and H. Raab in Austria. These observations are mainly of the "center" of the nuclear train, and this point continues to be the most relevant for orbit computations. Orbit solutions from positions of the brighter individual nuclei will be useful later on, but probably not until the best data can be collected together after the current opposition period.
>
> At the end of April, computations by both Nakano and the undersigned were beginning to indicate that the presumed encounter with Jupiter (cf. IAUC 5726, 5744) occurred during the first half of July 1992, and that there will be another close encounter with Jupiter around the end of July 1994. Computations from the May data confirm this conclusion, and the following result was derived by Nakano from 104 observations extending to May 18:

$$\text{Epoch} = 1993 \text{ June } 22.0 \text{ TT}$$

T = 1998 Apr. 5.7514 TT		Peri. = 22.9373
e = 0.065832		Node = 321.5182
q = 4.822184 AU		Incl. = 1.3498
a = 5.162007 AU	n = 0.0840381	P = 11.728 years[8]

> This particular computation indicates that the comet's minimum distance Delta_J from the center of Jupiter was 0.0008 AU [i.e., within the Roche limit] on 1992 July 8.8 UT and that Delta_J will be only 0.0003 AU (Jupiter's radius being 0.0005 AU) on 1994 July 25.4.[9]

PERIODIC COMET SHOEMAKER-LEVY 9 (1993e)

Following the remarks on IAUC 5800 concerning the encounter with Jupiter in July 1994, it should be noted that the 3-deg differ-

ence in p.a. [the position angle of the train of comet fragments] between the comet's direction of approach and the orientation of the
nuclear train should increase immediately before encounter, and the
undersigned's initial estimate is that more than half of the nuclear
train could collide with Jupiter—over an interval approaching three
days. . . . It must be emphasized that a 1994 collision of the train
center with Jupiter is not assured, and in the case of a miss, the 3-
deg difference in p.a. would minimize the chance of collision with
any part of the train.[10]

Even before these announcements, Paul Chodas and Donald
Yeomans of JPL were testing a program that had just been augmented, thanks in part to the recent Hazards Due to Comets and
Asteroids meeting held in Tucson that January, to predict the probability of impact of a comet or asteroid. After trying to create orbits
for theoretical objects that would lead to collisions, here was a real
test. Using an orbit based on more than one hundred positions of
the comet's center spread out over two months, many of these positions from Jim Scotti, Chodas, and Yeomans came up with a 50
percent chance of a collision.

The situation evolved very quickly after the May 22 announcement. As more astrometric positions of the comet were plugged
into the program in the days following, Chodas and Yeomans
raised the odds to 64 percent, and a week after that, to a virtual
certainty at 95 percent.[11]

FIGHTING FOR TIME

Suddenly realizing that he was overcommitted again, Gene and
Carolyn wondered if he had made the right decision to join the
Clementine team. Also, the Shoemakers' plan to end the observing
program at the end of 1993 was reversed by a NASA decision to
fund our expenses for a year, thereby allowing the program to
continue during 1994. With the Shoemakers planning to move to
Alexandria, home of Clementine's Batcave for the first four months
of 1994, working the Palomar mountain project was going to be
difficult.

When Comet Shoemaker-Levy 9 first came into our lives, it seemed as though it was just another comet. All that changed on May 22, the day the comet grabbed us and flew for more than half an orbit about Jupiter. Our time would increasingly be taken up with conferences, interviews, the writing of articles and papers, and conversations lasting late into the night as we tried to ponder the implications of this first collision of a comet against a planet ever observed by humanity.

THE CRASH BASH

On Monday, August 23, 1993, Gene and Carolyn joined some 120 scientists at a unique event called the Comet Pre-Crash Bash in Tucson. Both NASA and the National Science Foundation announced that funds would be made available for special impact-observing projects. Gene presented his informal study on the history of the comet's orbit, and noted that two other periodic comets, Gehrels 3 and Helin-Roman-Crockett, had made brief detours into orbits around Jupiter rather than the Sun. Although these comets had now abandoned that phase, both will be recaptured by Jupiter sometime in the future.[12] In order for such unusual cometary captures to take place, Gene noted, a comet needs to be circling the Sun in an orbit similar to that of Jupiter. It would enter a new orbit around Jupiter by passing through the Lagrangian or L1 point and later resuming its solar orbit by passing through L2. (In a system with a larger and a smaller body, like the Sun and Jupiter, there are five points, designated Lagrangian points or L1 to L5, in which a third body (like a comet), could remain fixed relative to the other two. L1 and L2 are unstable points, so a comet passing through them would not stay there. These gravitational effects were first noted by the astronomical dynamicist Joseph Louis Lagrange, in 1772.) Although it is not usual for Jupiter to affect the orbit of a comet in this way (perhaps once per century), it is rare for a comet thus affected to then proceed to enter the planet's Roche limit.

Shortly after that meeting ended, Gene and Carolyn escaped for a few month's R and R in Australia. They returned in mid-October, just in time for a dinner honoring Gene's retirement from the U.S. Geological Survey. Held at Flagstaff's Woodlands Plaza Hotel, the evening was a cross-celebration among geologists, astronomers, administrators, family, as well as a U.S. senator-and-astronaut. A favorite story that evening: a reminder that geologist Donald Gault had bet Gene that there would be a successful Moon landing by 1970. "OK," Gault noted, "so I lost, but only by six months. Just goes to show you how smart Gene is."[13] Another was the reminiscence by Senator-Astronaut Harrison "Jack" Schmitt. Schmitt began as a politician, congratulating Gene on his many years of service to geology, and after a sentence or two, he literally threw his notes away and turned to his friend, reminiscing about the unique experience Gene had given the astronauts. Finally, there was geologist Bevan French's song, specially written and performed for the occasion, *Just Passing By on My Way to the Moon.*

> He was born in a basin
> that's now called L.A.
> He decided quite early
> that he wouldn't stay.
> The Moon shone down on him,
> there were rocks all around
> And in that combination,
> his life's work was found.
>
> He'd lie in his cradle
> and smile at his mother,
> With a hammer in one hand
> And a rock in the other.
> And late in the evening,
> you might hear him croon
> "I'm just passing by
> on my way to the Moon."
>
> Chorus:
> He's done coesite and stishovite
> and asteroids and dinosaurs,
> He's discovered the craters

with which Earth is strewn.
He's done missions and
committees and management
and bureaucrats,
All the things that you do
on your way to the Moon.

He started with field work
Like all Survey hands,
But salt and uranium
were not in his plans.
It was Meteor Crater
and all of its kin
that changed our whole view
of the world that we're in.

Then he hooked up with NASA
And worked with *Apollo*,
'Cause where astronauts went,
geologists could follow.
And in conference or meeting,
he'd sing the same tune,
"I'm just passing by
on my way to the Moon."

Chorus:
He's done coesite and stishovite
and asteroids and dinosaurs,
He's discovered the craters
with which Earth is strewn.
And all these catastrophes
are non-Uniformitarian,
That's what you learn
on your way to the Moon.

So to all young geologists
who are new on the scene,
if you want to do well,
take your lessons from Gene.

Stay close to your field work
but leave your mind free,
And don't sit at home when
there're new worlds to see.

For the young are not finished
with the worlds that we know.
They've heard all our stories,
and they're eager to go.
It won't be next August
or the following June
But one day they'll pass by
on their way to the Moon.

Chorus:
He's done *Ranger* and *Surveyor*
and *Voyager* and *Clementine*,
He's explored, and he's taught,
and he won't slow down soon.
For in spite of committees and
all of those bureaucrats,
There'll be folks passing by
on their way to the Moon.

Reprise:
That's not bad for a man
on his way to the Moon.[14]

Another thought most of us shared: "Congratulations on your retirement," wrote Naomi and Jerry Wasserburg. "Now you can get on with the planetary travel that you really want to do. P.S. What retirement?"[15]

CHAPTER 20

Yes, Virginia, Comets Do Hit
Planets: 1994—E. M. Shoemaker, *April 1995*

How hard it is to hide the sparks of nature!

—Shakespeare, Cymbeline, *circa 1609*

Be not with mortal accidents oppress'd,

No care of yours it is, you know 'tis ours.

—*Jupiter's visitation in* Cymbeline

Throughout my career, I have dreamt of witnessing an impact event," Gene began in his address in Boulder. "Little did I think that event would happen on Jupiter."[1] Nature has a way of proving or disproving the ideas of mere mortals, but rarely has she shown a man's work to be right so spectacularly as during her splendid show on Jupiter in the summer of 1994. Like Jupiter's appearance in Shakespeare's *Cymbeline*, planet Jupiter pointed out how right Gene was, throughout his life, to focus on the role of impacts in the solar system.

After an informal gathering at the Shoemaker home the morning after Gene's retirement dinner, the Shoemakers and I left for Phoenix to catch a flight to Denver, and a rainy drive to the Boulder meeting of the Division for Planetary Sciences. The DPS is a part of the American Astronomical Society and the world's largest gathering of planetary astronomers. As we boarded the plane and *finally* relaxed, I casually asked Gene where the couple would be staying in Boulder.

"Well," Gene laughed, "we were going to talk with you about that!" I always admired his ability not to worry about littler details

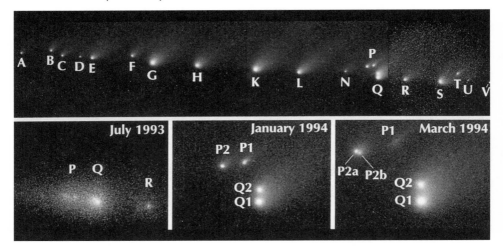

31. The Hubble Space Telescope's January 1994 view of S-L 9, six months before impact. NASA photo.

like keeping off the street, especially after they'd just finished traveling around the world. I suggested that if my friend Steve Edberg and I would share a room, then the fully booked hotel would have a room for them.

The "Boulder incident," I soon learned, was not atypical of Gene's way of traveling. During one 1960s flight, he stored his coat in the overhead bin—planes from that era did not have secured storage compartments. When his flight landed at Chicago O'Hare, Gene got off his plane and crossed the airport. When he got to the check-in counter for his continuing flight, he reached into his coat pocket for his ticket. It wasn't there. Everything—his ticket, his wallet, all his identification—had slid out of his coat pocket and was probably still in the overhead compartment of an aircraft that was, at that moment, rushing down the runway for some other city.

Gene wasn't sure how he'd get home. The agent, in this precomputer time, had no way to check whether anyone named Shoemaker was supposed to be on the flight. But Gene somehow managed to persuade the agent to give him a ticket, based only on a promise that he would submit a government travel request, or

GTR, to cover the cost of the ticket when he got home! The only thing he had to supply was his phone number. Of course, when he returned home he promptly forgot about the GTR until the airline telephoned to remind him of it.

Gene seldom remembered to bring money with him when he traveled. At a restaurant he would offer to pay the whole bill with his credit card if the others would pay him in cash, and thus, after the first meal he would build up his cash supply. He ended one particular trip without any money at all, and as he spent several hours in Albuquerque waiting for a delayed flight to Flagstaff, he found out that the airport restaurants did not take credit cards. Getting hungrier and more aggravated by the minute, he finally thought he recognized someone who lived in Flagstaff. Gene introduced himself, struck up a conversation, and borrowed ten dollars from the man for his dinner![2]

Gene usually managed to get his way, though gentle persuasion didn't always work. One night at Palomar, for example, the observatory front-gate area was crowded with people observing the 1992 Perseid meteor shower, which was expected to be an unusually strong one. We were following our regular program at the eighteen-inch. After the Moon rose we closed the telescope and set out to return to our quarters, but an old VW beetle blocked the entrance at the gate. Gene called out, but no one came up to claim the car. So he got into the car, put it in neutral, removed the brake, and moved it himself—a dangerous stunt in violence-prone California.

Gene was always in a hurry, even when he wasn't. During one trip he and Carolyn were trying to get to the gate for their connecting flight at Dallas–Fort Worth Airport. As they descended the escalator to the train, Gene saw the doors begin to close. Even though the next train was less than five minutes behind this one, he catapulted down the escalator and made the train just as the doors closed. "I just waved at him and laughed," Carolyn remembers. "The expression of absolute shock and disbelief on his face—that after all his racing, he didn't gain anything—was unforgettable."[3]

THE BOULDER MEETING

On the afternoon of Monday, October 18, Gene, Carolyn, and I moderated a two-hour paper session on Comet Shoemaker-Levy 9. The Hotel's grand ballroom was standing room only as Gene delivered his forty-minute talk about the comet's discovery and early implications. Then each of us chaired a third of the session that followed. It was a most fruitful afternoon, perhaps mostly for the list of key observations that were pointed out as needed to understand conditions on Jupiter *before* the first impact, and to know the impact times and places as far in advance as possible so that spacecraft like *Galileo*, and the Hubble Space Telescope (HST), could be scheduled accurately to catch them. At that meeting, Clark Chapman, a scientist with the Galileo mission, told Gene that the circumstances appeared to be unfavorable for observing from Earth—the orbits indicated that the comet fragments would approach Jupiter from the south, then swing behind the planet and strike on the far side.[4] The results of each hit would not be visible from Earth for at least an hour.

That evening Carl Sagan paid his greatest tribute to the comet, and to the man whose vision made possible its discovery: "This is a most extraordinary find," he said, and then described the circumstances:

> There are those whose idea of a good time is to stay up on cold nights taking pictures. They get enormous pleasure out of it: Eugene Shoemaker, Carolyn Shoemaker, and David Levy . . . have been at this before—Shoemaker-Levy 9 means that this is their ninth periodic comet find. . . .

Sagan then connected the S-L 9 event to an ancient impact on Earth:

> One hundred times ten to the sixth years ago, mammals were nocturnal insectivores, no larger than six inches wide. They were timid because the planet was owned by the dinosaurs. And yet, 65 times

10^6 years ago, something happened. . . . The dinosaurs were a victim of cosmic roulette.

His conclusion: The lesson of Shoemaker-Levy 9 was a serious and personal one: that if a civilization is to survive, it must have the ability to go into space and move threatening objects out of harms's way.

> We have to be in the world-moving business. Jupiter keeps comets away from Earth, but not all of them. In the long term, all civilizations must be spacefaring. The ones that aren't, die.[5]

While Sagan was talking, the comet of his focus was on the far side of the Sun, out of view of Earth. At the end of November, astronomer David Tholen used a telescope atop Hawaii's Mauna Kea to observe the comet as it appeared in the morning sky just before dawn. With the new astrometric measurements, the situation looked much better: the fragments would still swing around the back before impacting, but within a few minutes the damage, if any, would be visible from Earth.

Impact Year Begins

We began 1994 with a drive from Flagstaff to Palomar. Our first night was magnificently clear and warm. But as the nights passed, the temperature dropped to lows around thirty-four degrees. Staying outside for thirteen hours in that temperature was difficult; as Carolyn quipped, "our words fell on the observing floor and broke into a thousand pieces." On January 9, the day the observing run ended, we flew to Baltimore for a second planning meeting. Its purpose was to coordinate the most massive observing campaign to watch a single event since Galileo first turned a telescope to the sky in 1610. The crusade would include telescopes in most of the countries of Earth, telescopes orbiting the Earth, and spacecraft in several corners of the solar system.

As amateur astronomers used small telescopes to try to spot the impact flashes, the mighty Hubble Space Telescope would be watching as well. The *Galileo* spacecraft, then on its way to Jupiter, would be able to see the impacts directly, while those on Earth would not. Finally *Voyager 1* and *Voyager 2*, now far more distant than any of the planets, would also be looking back toward Jupiter.

IMPACTS PAST, AND IMPACTS PRESENT

At the end of January 1994, Gene and Carolyn watched a *Titan 2* rocket soar into the sky. Just as they had years earlier, the couple were watching a rocket head toward the Moon. This time, it was a young woman named Clementine who was, metaphorically at least, on her way to the Moon. The first mission to the Moon since Gene Cernan and Harrison Schmitt had done geology there twenty-two years earlier, the spacecraft named *Clementine* would map the Moon in ways that had never before been seen. As scientists prepared to observe a modern impact, *Clementine* was mapping a world filled with evidence of impacts past.

By mid-March, we were at Palomar again, with a new comet suspect but without Gene, who was locked in the Batcave in Alexandria monitoring *Clementine*'s seventy-day Moon-mapping spree. At the end of April, with impact less than three months away, I completed a lecture to the National Science Teachers Association at their meeting in Anaheim, California, and then met Carolyn at Orange County's John Wayne Airport. Happily, the rough travel didn't damage the stereomicroscope. We drove straight to Palomar and immediately took six photographs; the next day Carolyn began a scan of the first pair. "*Thar she blows!*" she said a few minutes later as she spotted the two trailed images of a fast-moving comet, soon to be called Shoemaker-Levy 1994d, or as Steve Edberg dubbed the racer, "Comet Speedster." It was amazing that we got any observing done at all during this period, for our observing runs were now accompanied by a cacophony of reporters from all over the world. At first reluctant to accommodate their needs, Palomar's management, realizing that their observatory was in the midst of a major news event, became very cooperative.

THE HURRICANE BEGINS

On May 10, 1994, a "convergence of the twain" took place over much of the United States as the Moon covered the Sun in an annular eclipse. Since the track passed through Las Cruces, I joined Pluto discoverer Clyde Tombaugh and Voyager team leader Brad Smith to watch the Moon pass over the center of the Sun, leaving only an outer gleaming ring of sunlight. Wendee Wallach, my future wife, was teaching classes in physical education at a nearby school and allowed her students to view the eclipse with protective glasses; the day before, I had lectured to the children about eclipses and impacts. As maximum eclipse ended, I left Las Cruces and drove for twelve hours to meet the Shoemakers who were on the first night of their May run. With the end of *Clementine*'s survey of the Moon, Gene was relieved to become a part of the observing team once again.

Since the observing time on the eighteen-inch telescope was divided equally between our group and Eleanor Helin's, on alternating months we would have the week before new Moon; this month we had the week after. With the eclipse marking the moment of new Moon, this was our first night. The next day a *Time* photographer arrived to photograph the Shoemakers, me, and the eighteen-inch. A few days after that, *Time*'s impact story appeared on the cover of their May 20 issue, and with it, the last vestige of our private lives ended for a few months. Our next observing run began on May 31; there was only a two-week break between the two sessions since we were assigned the first week of the dark-of-the-Moon period. We were joined by a camera crew from NBC who filmed us until 2:30 A.M. Meantime, with forty-six days until impact, our comet's first fragment was picking up speed as it silently approached Jupiter.

The following morning, a BBC crew filming two half-hour documentaries spent the day with us. On Friday, June 3, with forty-three days to go, a NASA camera crew filmed us at the telescope—their archival footage would appear on TV screens often in the coming weeks. The afternoon of June 5 was spent in another interview with the NBC crew. The next afternoon a group of young scouts paraded through our observatory, and on June 7, a TV crew from Mexico City filmed us. Despite this heavy bombardment of media, we still managed to enjoy a successful observing run with hundreds of photographic films. On Wednesday June 8, we drove home; thirty-eight days to go.

With excitement mounting, we were concerned that the event might be seen as a fizzle by the public. After all, a series of comet fragments not larger than villages was about to pound a planet that could hold thirteen hundred Earths. "Even if the Hubble Space Telescope doesn't see a thing," Carolyn assured me one night at Palomar, "this event will be a wonderful learning experience about the nature of impacts."

Sunday, June 26: With twenty days to go, at the Astronomical Society of the Pacific's Flagstaff meeting, Gene and I gave lectures to overflow audiences. By this time all three of us were targets of strange incidents. A senior police officer in New York telephoned Carolyn. Worried from articles in the tabloid press, he wanted to know whether he should fear the results of the impacts on his city. "He wanted to ensure that the people of New York would be safe— could I assure him they would be? I could indeed." People asked all three of us if we had read the biblical story in Revelations of mountains falling from the sky.

On July 8—just a week left—Gene, Carolyn, and Henry Holt took over the Palomar observing run while I was busy with almost nonstop interviews in Boston, New York, Philadelphia, and Atlanta. On the morning of Friday, July 15, the Shoemakers and I were reunited, electronically at least. I was back in New York in the studios of the *Today* show, and a camera crew filmed Gene and Carolyn live at Palomar Mountain. The next day, July 16, the three of us met amid a sea of media antennas and cameras at the Space Telescope Science Institute in Baltimore, just as fragment A, now hurtling along at sixty kilometers per second, approached its fated rendezvous with Jupiter. We were greeted by CNN reporter Miles O'Brien, who assured us that his network was devoting heavy coverage to the impacts, and that this had better be good! "Maybe we'd better find a flat rock to hind under," Gene quipped.

IMPACT

Now well inside Jupiter's magnetosphere, fragment A was being stretched as its forward portion experienced greater gravitational stress from Jupiter, and the Hubble Space Telescope imaged it one final time. At 4 P.M. Eastern time, the fragment, hurtling forward at sixty kilometers per second, struck Jupiter's outer atmosphere. With two orbits to go before data could be downloaded from the Hubble Space Telescope and viewed, we had little to do but wait for almost three hours. Reporters simply mobbed the main floor of the Space Telescope Science Institute. In a meeting room below, a group of scientists were finalizing their strategy. At about this same time Hal Weaver, the scientist in charge of HST's S-L 9 comet observation program, heard some news on the radio, saying "that something associated with the A impact had been seen at the Calar Alto Observatory in Spain. This was our first hint that this night was going to be special."

This was great news, but also frightening news. "This news from Calar Alto," Weaver continued, "only heightened my anxiety because I realized that these were infrared observations. There was a preliminary report from an optical observatory in the Canary Islands that nothing was seen. This could be our biggest nightmare: if the impacts were infrared-only phenomena, then we would be stuck trying to explain why the most expensive astronomical instrument ever built, the HST, couldn't see what even a moderate-sized ground-based observatory picked up! I couldn't let myself feel too good about these preliminary reports until I knew how HST had done, and I wouldn't know that for at least three more hours."[6]

Gene, Carolyn, and I were then ushered into the Institute's auditorium where NASA's Janet Ruff and Donald Savage were preparing us for the evening's press conference. We were each expected to give brief presentations—Gene on impacts, Carolyn on the discovery of S-L 9, and me on the importance of this particular event. As we were preparing, Hal Weaver walked in excitedly and whispered his news to Gene.

"You mean they saw a *plume*?" Gene bellowed. That put a stop to the rehearsal. We made a dash to a phone to confirm the report, first with a phone call to Brian Marsden, and then to my e-mail, where a dozen messages from Spain, Antarctica, and other places confirmed that a 3,000-kilometer-high plume, as tall as 360 Mt. Everests, had just risen over Jupiter.

When the press conference began that evening, the atmosphere was electric. Gene presented a model he had just completed with his colleagues David Roddy and Paul Hassig at the USGS. It showed a hypothetical plume from a one-kilometer-sized body rising over Jupiter's atmosphere, then collapsing on itself. If the reports from Calar Alto in Spain were right, exactly the same model was playing out over Jupiter right now.

A floor below, the scientists we had met earlier in their meeting room were now huddled around a single computer screen. "Heidi Hammel sat directly in front of the monitor where the images were to be displayed," Weaver recalls. "The rest of the team formed a circle looking over her shoulder. The first image of Jupiter (taken through a methane filter centered near 889 nm) showed a hint of something near the limb, but we thought that this could be an image artifact, like a cosmic-ray event. The second image, taken three minutes after the first, didn't seem to have anything unusual. But then the third image (again taken three minutes after the previous one; due to instrument operational overhead, consecutive images must be taken at least three minutes apart) looked very strange; there seemed to be a little ball slightly offset from Jupiter. Could this be a *Galilean* satellite? I quickly grabbed a nearby copy of the Astronomical Almanac and, with Melissa McGrath, verified that the ball could not be a satellite.

"After the next image the answer was clear; we were watching the development of a plume off the edge of the planet! For a moment the group of scientists just sat there, stunned. 'Oh,' Heidi Hammel exclaimed, 'My God!'

"We realized that we had something truly spectacular on our hands," Hal Weaver continued. "Melissa McGrath ran upstairs to get the champagne she had brought for the occasion (even though

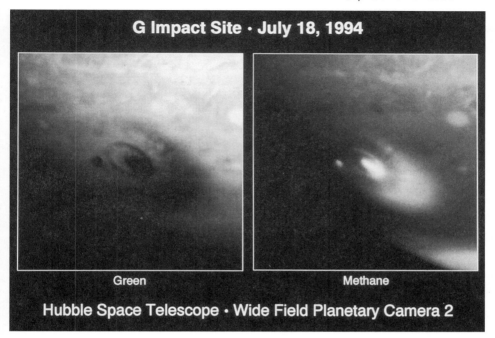

G Impact Site · July 18, 1994

Green

Methane

Hubble Space Telescope · Wide Field Planetary Camera 2

32. A startling cloud forms after the crash of S-L 9's fragment G, in this July 18, 1994, Hubble Space Telescope image. NASA photo.

she is the first to admit that she hadn't really expected to see anything like this), and Heidi Hammel popped the cork.

"The feeling of elation was indescribable," Weaver concludes, "and I doubt that I will ever experience anything like this again. This was not the 'Big Fizzle' that had been predicted only one week earlier,[7] but rather the most dazzling astronomical display of the century."[8]

A few minutes later, Heidi Hammel took the raw output of the image, walked upstairs, entered the auditorium, and interrupted our press conference with her stunning news: Hubble had imaged a plume (Gene raised his fist in triumph at that), and an hour later a growing dark cloud. Gene was ecstatic. "Those reports from Chile and Spain are *right*!" he exulted. But time was running on; we had an appointment with Fragment B, whose impact we hoped to see directly using the old 24-inch refractor at the U.S. Naval Observatory. Led by Gene, our three-car convoy sped along the highway and arrived at the Observatory. Although Jupiter was low and the air very turbulent, we did see the sphere of Jupiter through the eyepiece. We did not see any effect of the second impact.

Not surprising, that one was rather poor, and indeed a fizzle. Fragment B probably consisted of dust so loosely held together that it fell apart before impact. But the next day we heard about C, D, and E: "E was a dilly!" Gene said triumphantly. On Monday, July 18, the first reports filtered through of a stunning view of the largest fragment, G. The image from Australia's Siding Spring Observatory was so dramatic that it made almost every news program on Earth. It looked like an explosion, but it really marked the end of the event, as uncountable mass from the planet, and some from the comet, fell back onto Jupiter.

COMETOPOLITICS

By this time comet fragments were hitting everywhere, even onto the desk at CBS's *David Letterman Show*. The event was so extraordinary that millions of people turned from their daily lives to learn about the tribulations of a world 477 million miles away. Congress urged NASA to come up with a plan to locate all asteroids and comets a kilometer across or larger that could be a threat to Earth, and Gene was asked to chair the committee to develop a strategy. Gene spent most of July 18 huddled with his colleagues to get the group together and begin work.

The idea was fine in principle: take advantage of the momentum generated by media coverage of the Shoemaker-Levy 9 impacts, and expand the searches already in progress. But in practice, with congressional elections just three months away, the Shoemaker committee's work was cut out for it. That November vote, it turned out, changed the leadership in the House, and relegated the report to a back burner. Would the outcome have been different if someone with more time and focus had chaired the committee, making sure its report was done in time? The problem, several scientists commented, was that Gene couldn't say no. No matter how busy he was, he could not refuse taking on one more assignment. "I wouldn't put the blame on Gene," says Alan Harris, "I'd put the blame on whoever appointed Gene."[9]

By July 20, the excitement was starting to wear us down, but more was to come. In addition to being right in the middle of one of the most important weeks in astronomy since humans first looked toward the stars, this was also the twenty-fifth anniversary of the first manned landing on the Moon, an event that Gene had been heavily involved with as principal investigator for the geological field study. The early afternoon celebration was to be at the White House, and Gene, with the rest of the S-L 9 discovery team, was invited to participate.

This was the fourth time that Gene would get to meet a president of the United States, the previous event being his receiving the National Medal of Science from President George Bush in 1992. As this day's ceremony got under way, Vice President Gore introduced Neil Armstrong, the first person to set foot on another world. "Our old astrogeology mentor," the astronaut said, referring to Gene, "even called in one of his comets to celebrate the occasion with spectacular Jovian fireworks!"[10]

After meeting the President briefly, we were asked to wait by a member of the Vice President's staff: "If you have a few minutes, the Vice President wishes to speak with you." "Yes, we think we might have a few minutes," Gene laughed. Vice President Gore was in the middle of a radio interview, and once it was over he approached us. "Well," he smiled, "it's not often I get to meet three people as famous as you are this week!" The ensuing conversation was enlightening. Gore had already seen the impact spots from his home at the U.S. Naval Observatory, and he was as excited as any amateur astronomer who had seen the trauma that the impacts had inflicted on Jupiter. We all agreed with planetary scientist Clark Chapman's e-mailed comment from that day:

> I want to put this into the historical context of Jupiter observations. I have just come in from looking at Jupiter with my back yard telescope. The preceding end of impact site G is approximately on the

central meridian. Based on my own extensive experience of observing Jupiter when I was younger, and studying historical records of Jupiter observations from the early drawings of Hooke and Cassini through the extensive 19th and 20th century reports of the British Astronomical Association, I would assert:

THIS IS THE MOST VISUALLY PROMINENT DISCRETE SPOT EVER OBSERVED ON JUPITER.[11]

A TUCKS-NEEDING TUX

After being chauffeur-driven to the White House from Goddard Space Flight Center that morning, we were left to our own devices after the White House ceremony. We hailed a cab in the summer heat and humidity to take us to the Mayflower Hotel, where I was staying and where the three of us were invited to a Planetary Society formal dinner that evening.

The dinner was supposed to be incredible, with astronauts and cosmonauts, scientists and writers, and Vice President Gore as speaker. After the events of the day, this seemed just the ticket for some relative relaxation. Gene and Carolyn were staying in a hotel in Baltimore, so Carolyn and Gene used the rooms that my mother and I had in order to change for the dinner. As Gene was donning his tux, I returned a telephone call from a Phoenix radio station. As I answered the interviewer's questions, Gene started muttering "Damn! What am I going to do?" I asked the interviewer to hold on while I looked.

There stood Dr. Gene Shoemaker, the man who had just been congratulated by the President and by the first man to land on the Moon—and the tailor's pins were still in place in his tuxedo. He looked utterly lost. "Gene," I said helpfully, "I'll be with you in a minute." I went back to my interview, but a minute later Gene started talking again. "I don't believe it! I just don't believe it!" I turned around and saw the father of astrogeology wearing trousers that stretched several inches beyond the points of his shoes.

There was only one thing to do; I cut the interview short to deal with this turn of events. The Shoemakers had rented the tux from a Flagstaff outfit, so we had to get some local help. I fetched Carolyn and my mother. Soon the hotel's valet was trying to do emergency repairs. "Dr. Shoemaker," the valet said as she looked at the casual brown walking shoes, then up toward Gene, "I would be able to measure your trousers better if you put on your dress shoes."

Getting more exasperated by the minute, the geologist looked down at his boots, and then at the valet. "These are my dress shoes," he said.

Eventually the valet got the tuxedo to fit, and with borrowed shoes, Gene was ready for dinner. I returned the call to Phoenix and finished the interview, and then we went downstairs for dinner. On a "normal" day, the banquet would have been a fitting climax, but not for us. We stayed long enough to hear the Vice President's speech but then we had to run, tuxedos and all, to the hotel's second-floor lounge. There we found a full camera crew set up for a live, nationally televised program on PBS that featured Arthur C. Clarke from his home in Sri Lanka, as well as Gene, Carolyn, and me. The evening eventually closed with an almost unending line of autograph seekers.

FAREWELL TO A STRING OF PEARLS

In this way, impact week continued. On Friday morning, the last of the string of pearls, as astronomer David Jewitt had dubbed the comet, collided into Jupiter. As the fragment exploded in Jupiter's atmosphere, the event was captured, simultaneously and coincidentally, by two spacecraft from different vantage points in the solar system, the *Galileo* spacecraft and the Hubble Space Telescope. That evening we were honored on ABC News as Persons of the Week. "We come face-to-face with the tragedy of Rwanda," anchor Peter Jennings began, "but our attention has occasionally been diverted by what is happening in outer space. A disintegrating comet on the planet Jupiter has been something to behold. . . . In its death throes, it was spectacular."

Gene explained that Carolyn had "shed a few tears for some of the first pearls that went in."

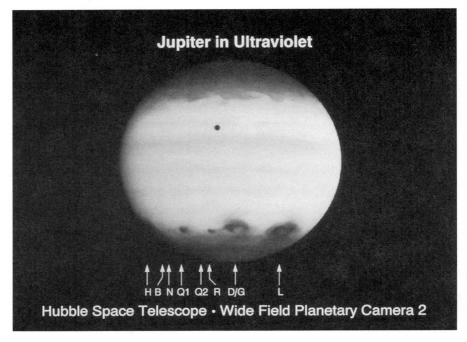

Jupiter in Ultraviolet

H B N Q1 Q2 R D/G   L

Hubble Space Telescope • Wide Field Planetary Camera 2

33. Several Earth-sized clouds remain on Jupiter's southern hemisphere after the impacts of S-L 9.

"And Gene, for my birthday," Carolyn interrupted, "gave me this nice string of pearls to replace the one I'm losing."[12] Despite the framing of national television, it was a touching moment.

Finally, the impacts were over, and the media hurricane moved past us. My mother returned to Montreal, I flew to Tucson, and Gene and Carolyn went to the serenity of their Flagstaff home. But Jupiter was still calling, and the next evening, the Shoemakers quietly drove to Lowell Observatory, whose old, large Alvan Clark telescope was focused on Jupiter. There, silently, they looked at the giant world, battered and blackened by the clouds left from their dying comet. "I've had this fantasy of seeing an impact," Gene had admitted on ABC news. Finally, on this quiet night at home, he realized that his wish had been granted.

CHAPTER 21

New Challenges: 1995–1997

When clouds are seen, wise men put on their cloaks;

When great leaves fall, then winter is at hand;

When the sun sets, who doth not look for night?

—SHAKESPEARE, Richard III, *1592*

WITH THE END of the impacts of Comet S-L 9, Gene knew that his life would face new opportunities, tasks, and challenges. One of these was clearly an increase in public appearances. People who know Gene Shoemaker well might be surprised to learn that as a college student, he was so wary of having to give a public lecture that he would shake visibly when lecturing to his own classmates at Caltech. As with many other challenges he faced in life, Gene met it as well as he could: to build confidence as a lecturer, he took a course in public speaking. Over time, Carolyn reports, he became much more comfortable lecturing in public forums or talking to the press, but some vestige of stage fright lingered throughout his life. At the lecture he gave about impacts at the Flagstaff meeting in June 1994, for example, I noticed him fiddling with the substantial number of keys and coins he had in his pocket. And Carolyn, who often sat in the back of the room to get a sense of the audience's reaction to his words, often noted that he would look at the floor when he spoke, instead of at his audience. We both agreed that looking at the floor was less a result of stress than a result of his focusing on his topic. "He really concentrates," Carolyn noted. "You can see and hear the wheels moving."

Gene became so comfortable with newspaper reporters that he could handle telephone interviews with them at any hour of the

34. Carolyn, Gene, and David whoop it up after the Planetary Society Jupiter Impact event in Washington, July 22, 1994. Photo by Bob Summerfield.

day or night. One reporter's phone call came very early in the morning, at a time of the night when Gene would normally be in his deepest sleep. He gave the interview, Carolyn remembers, with all his usual enthusiasm, using full sentences and colorful analogies. When the interview ended, he hung up the phone and fell right back to sleep.

In later years, Gene was comfortable in public, whether with reporters or with colleagues; in fact his ability to give unusual introductions to other speakers is legendary. People would be sure not to miss colloquia when they knew he would be *introducing* the speaker, for the background information he provided was so thorough and interesting. Once, however, his enthusiasm got the better of him. When Gordon Swann invited Walter Alvarez, the geologist who had led the team that discovered the iridium layer marking the Cretaceous-Tertiary boundary, to give a colloquium at the USGS in Flagstaff, he asked Gene to introduce the famous geologist. Gene hesitated. "The last time I introduced Walter," he laughed, "I got carried away and told his whole story!" But Gene did offer to do a short introduction. A few days later he prepared a cartoon slide of a dinosaur, and when Walter Alvarez arrived, Gene inserted his slide into the projector at the position he thought was the beginning of the Alvarez lecture. When the colloquium started, Gene gave an appropriately short introduction, and then asked for the slide. But the slide was in the wrong place. After picture after picture got shown, they finally found Gene's dinosaur image. At last, it was Alvarez's turn. "Last time Gene gave half my talk," he began, "this time he showed half my slides!"[1]

PALOMAR PROGRAM ENDS—ALMOST

Our observing sessions resumed in September, and the Shoemaker program continued observing through the fall of 1994. The night of September 18 was clear, and we took films until dawn, then hurried back to our lodgings for some sleep. Not much sleep that morning, though, for by nine-thirty we had met Bob Thicksten, observatory manager, who drove us and our friend Jean Mueller down to San Diego's City Hall. That was the day that the City Council proclaimed as "Palomar Observatory Day," "Eugene and Carolyn Shoemaker Day," and "David H. Levy Day" in celebration of the observatory's role in the discovery of our comet. Considering how accommodating the observatory had been throughout the sixteen months from discovery to impact, we thought that the celebration was an appropriate one for the observatory. With the December observing run—the last one before we packed up our tents—Gene threw a party for the entire staff at the observatory. Held at noon, it was a bittersweet time, for it marked the end, for Gene, of more than twenty years of observing at that mountaintop. Gene felt that with the advent of new electronic searching programs using charge-coupled-device (CCD) technology, we had been "beaten at our own game" and the time had come to stop.

Changes were happening in Carolyn's life, too. During the time after the S-L 9 impacts, she found herself working with Gene in his geological work even more closely than before. With the end of the observing, she had more free time to computerize his papers. Since Gene had an uneasiness with computers, this was a natural role for Carolyn, and in this way she became more a part of his research. Carolyn's growing experience in using word processing to write papers and proposals stood her well, despite a problem they had while in Australia in 1996: After spending many hours typing up an important paper on the significance of the tektites there, a malfunction seemed to send all her efforts to RAM heaven. Fortunately effective repair work from a friend brought the machine, and the lost paper, back to Earth. It's a good thing that happened, for the paper illuminates Gene's idea that the impact that produced the field of tektites was a major one that occurred less than three-quarter million years ago.

SHOEMAKER-LEVY 9: THE REPRISE

Throughout the late summer and fall of 1994, Gene's attention was still very much focused on the S-L 9 impacts. Demands for talks were way up for all three of us. Gene and Carolyn mastered their "Dog and Pony Show," where Carolyn would talk specifically about the discovery and fate of the comet, and Gene would follow with his views about impacts. For one lecture, they traveled with enough materials for one half-hour presentation each; but they were chagrined on learning that their time was cut by half. Their talks included the latest results from the impact, which meant that they changed with time. To keep up with every scientific aspect of the Shoemaker-Levy 9 story was virtually impossible, since it involved everything from plasma physics to the detection of new chemical species in Jupiter's atmosphere. Gene was particularly interested in several major aspects of the collision.

Size of the Progenitor

How large the progenitor comet was (before it broke apart on July 7, 1992) is still controversial. Gene was persuaded by a model by

Eric Asphaug and Willy Benz that suggested a relatively small progenitor comet.[2] Physicist Mordecai-Mark Mac Low saw the luminosity of the fireballs as indicating that the fragments had broken up very high in Jupiter's atmosphere, and he interpreted the chemistry of the impact spots and the lack of any seismic waves from the impacts as indicating that the largest fragments were not greater than 3/4 kilometer in diameter and that the progenitor was about 1.5 kilometers across.[3]

Gene and I had long debates about this question. Gene agreed with Clark Chapman's idea that small is beautiful. For comet fragments that small to have such a big effect on Jupiter's atmosphere, it is likely that asteroid and comet strikes on Earth were more devastating than we had thought. When Gene and I argued about it, I took the view of Zdenek Sekanina, Paul Chodas, and Donald Yeomans, who concluded that the way the comet train was angled after its discovery in March 1993 could not be explained if the comet had split up exactly as it passed perijove, or its closest point to Jupiter, almost nine months earlier. If they are right, S-L 9 was about ten kilometers in diameter when it broke apart due to tidal interactions with Jupiter.[4] Jovian tidal forces caused the comet to crack at first, and then it split along the planes of the tidally induced cracks due to the comet's rapid rotation. Some fragments split again later.[5]

History

S-L 9 began its wanderings, Gene thought, in the Kuiper Belt just beyond Neptune at the dawn of the solar system. Its capture by Jupiter began as a gradual process of several close encounters with Jupiter that gradually shortened its orbital period from several thousand to several hundred years. In about 1929, according to Chodas and Yeomans, the comet had a low-velocity encounter with Jupiter that allowed the planet to capture the comet as a loosely bound moon; instead of orbiting the Sun, the comet was now orbiting Jupiter in an unstable path wherein each revolution differed from the one before.[6]

Chemistry

Jupiter was still showing considerable changes wrought by the impacts more than two years after the last fragment disappeared in the planet's clouds. There was hydrogen cyanide (HCN) and carbon sulfide (CS), compounds that had never before been detected in Jupiter's atmosphere.[7] What was the origin of these chemicals? Since the high temperatures of the fireballs would have destroyed most of the organic materials, Peter Wilson and Carl Sagan suggested that they were formed by the process of "quench synthesis" as cometary material destroyed in the hot fireball recombined into the new chemicals and the temperature plunged in the plume while rising over Jupiter's atmosphere.[8]

Frequency

Gene was particularly interested in the frequency of S-L 9–type bombardments, since this directly related to his own life's research on impacts in the solar system. He suggested that a 1.5-kilometer diameter comet should strike Jupiter ever century or so, and that a comet striking while orbiting Jupiter would occur half that often. An S-L 9–type event, where a comet breaks up while in orbit about Jupiter, and its fragments later crash, might occur every two thousand years.[9] M. S. Roulston and Tom Ahrens thought that the event was rarer still, that comets in the kilometer size range of a single fragment of S-L 9 might strike at intervals of five hundred years, while comets as large as the progenitor comet collide with Jupiter only once each six thousand years.[10]

Gene ended his summary at the Baltimore meeting with these words:

> Now what are the odds to have such a rare event happen in the decade that all the new infrared detectors became available, as the *Galileo* spacecraft was in position to see the hits directly, and only six months after the Hubble was fully operational—*and* before the expected cutbacks make the money run out?
>
> Folks, we had a bloody miracle.[11]

A Trip to Saudi Arabia

At the end of 1994 geologist Jeff Wynn, a geophysicist with the USGS, invited Gene to join an expedition to map the Wabar impact structure in Saudi Arabia, arranged by the Zahid Tractor company to demonstrate the use of "Hum Vs" for travel over desert sand dunes. These structures are likely the result of the impact of a 300-ton body at least four meters wide that took place between one and six centuries ago. This structure was unique: the entire evidence for impact existed in the sands, which were utterly transformed into impact glass.

The expedition left in March 1995 and consisted of a rough ten-day trek deep into Saudi Arabia's bristling-hot "Empty Quarter" and the completion of the first detailed maps of the entire impact complex. They found three craters, beautifully exposed in the sand—Philby A and B, as they called them, were over 100 meters and about 650 meters across, respectively, and a smaller third crater, which Gene proposed be named for Wynn, stretched out at 11 meters.[12] With temperatures soaring past 140 degrees Fahrenheit, Gene worried about his salt intake, for a loss of salt at that temperature could cause him to become light-headed. He did have the presence of mind, and the enthusiasm, to want to collect almost every sample he saw—the sand impact was something completely new for Gene. He also met his match with Jeff Wynn, a geologist as animated, witty, and as totally devoted to his work as Gene was.

In March 1996 the Shoemakers added a second trip to Australia, this time to catch the southern coast while it was not suffering from near constant winter rains. This trip gave the couple the opportunity to join forces again with Ralph Uhlherr, the man who knew where the tektites were.

A New Generation of Observing

In a sense, our Palomar program marked an end to photographic surveying, an activity that lasted almost a century and peaked in

1930 (at least in the public sense) when Clyde Tombaugh discovered the ninth planet, Pluto, using two photographic plates. In the last years of our program, the future was building, and it would not be a photographic one. Instead it would center around the development, in the 1970s of electronic charge-coupled devices, or CCDs. These computerized chips were capable of recording images far more efficiently than film was, but until the mid 1980s the chips were not big enough to record the large areas of sky that survey work required. At that time Project Spacewatch at the University of Arizona began scanning with a CCD. By the early 1990s Gene was working with Ted Bowell of Flagstaff's Lowell Observatory to launch Project LONEOS, or Lowell Observatory's Near-Earth-Object Search. "Gene realized that with LONEOS, we would be beaten at our own game," Carolyn explained, "but with these new surveys, we would find out what's out there faster and more efficiently, and this was the thing Gene wanted all along." He looked forward to the start of the new program and planned to do some limited observing with it.[13]

THE SHOEMAKER-LEVY DOUBLE COMETOGRAPH

Late in 1994 Gene and I were minding the telescope at Palomar—Gene was guiding on a star, and I was marking time and ready with the next film holder. As we learned in chapter 1, Gene and I often had our best discussions during the four or five minutes between the time I came up from the darkroom with new film and the end of the exposure. On this night, though, we planned a whole new observing program.

"We need to find a way for Carolyn to keep finding comets," Gene began. With our program just two months away from ending at Palomar, I agreed. Gene thought he had a line on an unused Schmidt camera in Australia and suggested that we could ship the telescope across the Pacific. We could then set it up somewhere in Flagstaff. I reminded Gene of my two eight-inch f/1.5 Schmidt cameras. He called them "the babies" when he had first seen them some years earlier. As he pressed a guide button, he nodded as the telescope twitched in response. Before the end of the exposure, the Shoemaker-Levy Double Cometograph was born, and when Carolyn heard about what the conversation was all about a minute later, she was ready to start as well.

Planning the observing site was our next priority and occupied our conversation during the next few exposures. "We need a full dome in Flagstaff," Gene insisted. "It gets too cold there in winter, and we can't be observing bare-assed to the wind!" In the spring of 1995 we ordered a ten-foot-diameter "Home Dome." When the parts arrived in the fall, we began construction. Gene also arranged with Bob Thicksten at Palomar to give us the loan of a hypering oven. Now our team was complemented by Wendee Wallach, its newest member. On a much colder weekend a month later we four completed the dome. A month after that we aligned the telescopes on the pole.

We thought we were ready for photography, but over several months we ran into problem after problem, first with the focus on both telescopes, then with a mysterious fogging on our films. Gene used his considerable knowledge of mapping to prepare a tentative series of fields that we could observe two at a time, with both telescopes simultaneously! By June 1997 we were finally into the operational phase of our program. Gene, Carolyn, Wendee, and I completed some sets of fields, and then we developed the films in a darkroom reserved for us at the USGS. The program was running smooth as silk, although clouds had come in and rain was falling. Wendee volunteered to watch the films while Gene and Carolyn ran back to close their offices and fetch their car. Since this was a weekend, the Survey building was locked, and Gene had carefully explained the procedure for disarming the alarm so as to get in. However, he had not said what to do to get out of the building. The Shoemakers drove to the front door, and Wendee held it open, with the box of films still inside. Suddenly an excruciatingly loud alarm went off. Gene knew that Wendee had only a few seconds to disarm it before the building security guards were summoned. Waving his arms frantically, he shouted a list of complicated instructions to Wendee amid the pelting rain and thunder. Not hearing a word he said, Wendee looked back at him forlornly. "I don't know what to do!" she mimed. As Carolyn made the call to the security office, Gene started to explain how to disarm the system, but suddenly he dissolved in laughter. For some time afterward, Gene would smile and laugh whenever he remembered Wendee's look of absolute bewilderment as she stood out there in the rain.

Gene was also delighted that this new project was finally getting off the ground and producing consistent searches of the night sky for comets, and that the only remaining problems were in locked doors and alarms rather than in the more basic items of the program. In the years to come, the Shoemaker-Levy Double Cometograph project would continue to build on Gene's prescription: Cover as much sky as you can, keep good records and statistics, take advantage of breaks in the clouds, and never, never give up.

CHAPTER 22

Dr. Shoemaker, I Presume: 1997

Gene had a relaxed smile that would put you in a very

nice state. He made you feel that you had nothing to fear.

—PETER MARSH, *interview, 1999*

'A was a man, take him for all in all,

I shall not look upon his like again.

—SHAKESPEARE, Hamlet, *1600.*

The *Gods Must Be Crazy* is a witty and clever story about a native of Africa who encounters a fragment of modern civilization—an empty soda bottle—and roams far and wide to find and return it to where it came from. The movie satirizes the complexity of modern civilization, pitting it against a more basic form of life. Gene and Carolyn related to that dichotomy when they saw the film in February of 1986. They understood and loved its message. For three months of each year, they abandoned the soda-bottle complexity of their lives in order to spend time in a distant land, doing the work they loved. Seeing the movie at all was an uncommon act for Gene, whose life was his work. However, he did have a side that longed for rest and family. Faced with a lot of work to do, he once stopped, closed his folder, looked up at Carolyn, and said, "How would you like to go on a picnic? Let's go someplace we've never been before." These moments were rare, but they did happen, and sometimes they included movies. Gene did not generally like movies, or anything fictitious. He did enjoy reading biographies of the central scientific figures of our time, like fellow Princetonite Albert Einstein. "Or on long airplane rides," Carolyn reminisces, "he would often want to read whatever paperback I was in the middle of!"

Over the years, one of Gene's most successful ways of relaxation was designing, building, and finally living in his home. The construction took almost two decades. "He took great pride in our home," says Carolyn. "He had ideas for it since he was a young man." He at first envisaged a southwest style for the house, but when that turned out not to be practical in the Flagstaff climate, he designed a rock house, an obvious choice for a geologist. As building commenced and stretched over almost twenty years, visitors would find themselves relegated to helping unload materials, and because of the location of the house in the forest, heavy equipment could not get close enough to raise major structural elements like the logs that would soar majestically above the floor, or to raise heavy stones for the fireplace. To raise these beams, the Shoemakers organized work parties for students and colleagues, with a keg of beer available for all, as well as a home-cooked dinner. Once the house was completed, Gene looked forward to returning there after a day at the USGS, or especially after a trip. Gene lost some interest in the finishing details, like laying sandstone embedded with fossil footprints for the fireplace hearth. "But Gene and I loved the views," Carolyn explained, "and that's what the house was all about. It was Gene's castle."[1]

A Final Session at Palomar

Early in 1996 Gene excitedly told me that a producer for National Geographic Television, Eitan Weinreich, asked us to spend a day or two at Palomar so that they could film our work. "I thought we could turn this into a whole damned observing run!" Gene added as we planned this first-ever observing experience for the new Shoemaker-Levy team: Gene, Carolyn, Wendee Wallach, and me. We were excited at the prospect of having one more Palomar run; we had rather missed the experience since we had stopped more than a year earlier. The weather, it turned out, was fabulous, and we covered a lot of sky. For the second and third nights, Gene and

Carolyn bolted and went to Washington where Gene was awarded NASA's Exceptional Scientific Achievement Medal. During those nights Wendee and I observed solo on the telescope, and Wendee did very well indeed, especially considering that it was only the second night she'd spent doing astrophotography in her entire life! When Gene and Carolyn returned, he naturally went right to our films, and developed them. "A couple are a little dark," Gene typically *had* to comment.

A month later, we met again in Meudon, near Paris, for an International Conference on S-L 9. The meeting was especially enjoyable, since Gene and Carolyn were a part of our engagement announcement that week. They chose an asteroid that we could name after Wendee, one that we found on October 25 (coincidentally her father's birthday), 1991. Thanks to Brian Marsden and the Small Bodies Names Committee, it was all set to announce when I proposed to her in Paris.[2] A few days later, Gene and Carolyn, Clark and Lynda Chapman, and Wendee and I had a small celebration of life on a Bateau Mouche; we dined as the boat maneuvered down—and up—the Seine. It was a fine gathering of old friends. I especially appreciated Clark's presence, since he had encouraged me to meet Gene just before our first encounter in 1988. After that conference, the Shoemakers went on to a second assembly at Versailles, two meetings in Italy, and then spent a busy summer in Australia.

HOPES AND TRAVEL

On March 24, 1997, the day after Wendee and I were married, Gene, Carolyn, Wendee, and I met with David Taylor and Martyn Ives from the English production firm York Films. The aim of the three-day meeting was to prepare an outline for a proposal for a four-part TV special called *Origins*. A program of beginnings, it was also a new start for Gene, who planned to give up some of his science work in order to participate. The program would likely have taken up to two years of Gene's time and would have involved travel, narration on television, and writing.[3] We spent three days talking about the origins of everything from the galaxy, to the solar system, to the Earth, and finally to the appearance of humanity on Earth. Gene and I painted with a very wide brush that week as we pondered a succession of events that led us from the Big Bang, to the evolution of our Milky Way galaxy, to the giant molecular clouds that sat around in space waiting for something to happen. Something did: a supernova exploded, infusing the molecular cloud with organic materials and causing a portion of it to drift apart and start condensing. Events happened relatively quickly after that, Gene said. Moving his arms to simulate the evolution of the early system, he described how within 100 million years the cloud had condensed and its center had become hot enough to result in the ignition of the Sun. He spoke of the primordial Earth, adrift in a solar system filled with comets whose collisions with our planet brought the organic materials that were the building blocks of life.

It was when we talked about the origin of *Homo sapiens*, however, that Gene was entering ground I didn't know he was familiar with. He described some of humanity's earliest tools, ancient fishing harpoons, that dated back a hundred thousand years, and artifacts three times older than that—from Siberia. He talked about how Neanderthal man coexisted with Cro-Magnon man some thirty-thousand years ago in what is now the land of Israel.

Coming from a man who traveled most of his life, Gene's assertion that Cro-Magnon man "was built to travel" especially made sense. Travel, Gene knew, meant flexibility. While Neanderthal succeeded in his own local area, Cro-Magnon was able to move about and establish relations with his neighbors that Gene believed produced the cultural evolution that gave him an edge. With travel, Cro-Magnon and his descendants, *Homo sapiens*, spread throughout the world—out of Africa by about 120,000 years ago. By sixty thousand years ago, modern humans were thriving in Australia. "They pretty well romped across southern Asia and into the Indonesian Archipelago, and during glacial maxima, they could use land bridges," said Gene. "But how did they get to Australia? Once you got to New Guinea they could walk to Australia, but it's not quite clear how they got to New Guinea."[4]

As our conversation went on, Gene expounded his own pet theory about the rise of agriculture. He thought that it coincided with climate shifts. "The period of hunter-gatherer dried up when the Sahara was still a steppe," Gene maintained, "but a dramatic shift in climate forced people into a new mode of substinence. The disappearance of the ice sheets and the drying up of the land forced the development of agriculture along the Nile, the Tigris, and the Euphrates."

Simply listening to Gene's ideas on so many subjects, his wide range of interests, was a phenomenal wedding gift from him. After the meeting broke up we all looked forward to the program that would surely follow. "Everything is going to start to happen," Wendee said as the three days of meetings ended and the tape turned off.

THOUGHTS OF A FUTURE

Late one night at Palomar, Gene and I talked about a friend I had lost in a small plane crash. Gene commented that when he died he wanted to die quickly. "I want to go with my field boots on," he said, "like falling off a cliff." Gene had the chance to reflect on death again in the spring of 1997, when Jurgen Rahe, a well-known planetary scientist at NASA, was killed when a tree fell over his car while he was driving home in a Washington thunderstorm. In mourning the loss of his good friend, Gene explained to his family that when that time came he wanted to die suddenly. In the back of his mind was the way his own father had died almost forty years earlier.[5]

All this time, a spacecraft named *NEAR*, for Near Earth Asteroid Rendezvous, was approaching its encounter with the asteroid Mathilde, the first-ever encounter with an asteroid on a spacecraft designed and dedicated to the study of asteroids. (In 2000, the craft was renamed NEAR-Shoemaker.) When the event took place in June 1997, Gene and Carolyn were part of it. The asteroid was heavily cratered and it appeared as though major impacts had

35. Carolyn and Gene, and Lynda and Clark Chapman dance a *hora* at the Levy wedding on March 23, 1997, exactly four years after the discovery plates for Comet S-L 9 were taken. Note the "Colliding Asteroids" display above, at Flandrau Science Center. Photo by Bob Summerfield.

carved pieces out of it. "Gene looked at the pictures at the same time he was getting information on the size and mass of the asteroid from *NEAR*'s radio tracking experiment," Carolyn Porco describes. "Then Gene went to his hand calculator, and in thirty seconds he pronounced that this asteroid must be a sand pile! He had that kind of facility with cratering mechanics and the structure of bodies like asteroids; in no time at all he came to this fundamental conclusion about what Mathilde was all about. He did this in real time with his hand calculator."[6]

At the same time, dynamicist Donald Yeomans was making an initial calculation on the asteroid's density; he concluded that it was about 1.3 (1.3 times denser than water). Though Gene believed this result, his colleague Al Harris thought it was too low. Gene said, "OK! I'll bet you!" He bet that it was less than 1.5, and Harris that it was denser than 1.5. Gene won: the final density result came in at about 1.4.[7]

Flushed with excitement at the encounter with Mathilde, the Shoemakers flew directly to Montreal, where they met Wendee and me to start a lecture trip in Canada. We took a train to Kingston, where Gene presented his Ruth Northcott Memorial Lecture on Impacts. Named after one of Canada's best-known astronomers, these lectures are given every two years at national meetings of the Royal Astronomical Society of Canada. The previous two had honored the Shoemaker-Levy 9's discovery team. In 1993 I talked in Halifax about comets, and two years later Carolyn spoke in Windsor about the Palomar program. Now it was her husband's turn.

Gene's lecture was a beautiful summary of his life's work, and it emphasized his research in cratering on the Moon, in Australia, young features like Meteor Crater and Wabar, and comets. He ended by suggesting that the 26- to 35-million-year-period between comet perturbations in the outer solar system is a real one, and that we are "very close to a plane crossing now." His evidence: the Helium isotope ^{3}He, once thought to be the product of volcanic eruptions, may actually be of interplanetary origin, having been brought to Earth by dust particles from comets. Gene believed that the amount of this isotope on Earth peaks every 30 million years

36. Stefani Salazar combing her grandfather's hair.

or so, and may be related to comet showers that occurred in the past. The dinosaur extinction 65 million years ago and impacts at Chesapeake Bay and in Russia 35 million years ago may be related to a pattern of periodic comet showers. Gene ended with a bombshell: we may be just coming out of a comet shower now; a period that began two million years ago is just ending. The helium evidence, as well as the tektite glass he and Carolyn had been examining in Australia, made Gene believe that a major impact may have taken place some three quarters of a million years ago and that the crater is waiting to be found in land or in shallow water near

Indochina. "Given enough time to search," he said, "we should find it." What caused this periodic flux in the numbers of comets blanketing our sky? Gene suspected that as the solar system pushes its way, like a sine wave, up and down in the galaxy, it encounters the dense plane of our galaxy every 30 million years. The increased gravitational disturbances that are related to the plane crossing may increase the chances that Earth, and the other planets, get struck by comets that have been perturbed into the inner part of the solar system. Ending his presentation with a grand flourish, he concluded thus: "The study of the impact history of Earth is in reality a way to study the history of the motion of the Earth in the galaxy."[8]

July 18, 1997

Our trip continued to the town of Wolfville, Nova Scotia, home of the highest tides on Earth, and of my alma mater Acadia University. Gene, Carolyn and I each presented a lecture on that evening of July 2, and the next day we flew home. As we parted at the Toronto airport, we hugged and made plans to meet again at the end of the summer for our next observing session. A week later, Gene and Carolyn arrived in Melbourne and visited their friends Ralph and Herta Uhlherr, to discuss their search for tektites. They drove to Adelaide, where they picked up their Toyota Hilux. As they headed north, they felt absolutely wonderful about being back in Australia. Neither Gene nor Carolyn felt the need to wear seat belts. Their immediate agenda: to revisit Goat Paddock, an impact structure gouged less than 50 million years ago.

On the eighteenth of July the couple left Alice Springs and drove until it was almost dark. "We camped in a pretty desert area for our first night," Carolyn says. "I tried to see Comet Hale-Bopp in the early morning hours. I thought I saw it and woke Gene."

"Carolyn, do you have your glasses on?" Gene chuckled, as he confirmed that it wasn't the comet his wife was looking at. The next morning, July 18, Gene and Carolyn continued their drive along an arrow-straight road. On occasion they would see the dust

37. Wendee and David Levy with Gene and Carolyn, after their car broke down on the way to a lecture in Wolfville, Nova Scotia, July 1, 1997. We rented a limousine, and made the event in style! Roy Bishop photo.

cloud of an approaching vehicle more than a mile away. They passed one gold mine, then another. Several trucks passed, which was rather unusual; often they would pass only a single vehicle in a whole day. Once they passed the mines, however, they didn't expect to meet anything else. They stopped briefly to transfer gas from one of their fuel drums.

It was almost one o'clock. As the couple drove along the straight road, they looked forward to visiting with Dan Milton, one of the first geologists to take an interest in Australia's impact history, that evening at Hall's Creek. Their conversation was relaxed; Carolyn asked where in the town they were supposed to meet him. "It's not such a big place," Gene answered. Then he grinned as he reassured his wife, "We'll find him." Driving in the center of the corrugated road, Gene and Carolyn were doing what drivers are expected to do in the outback. Gene was in the driver's seat on the right side of the car; if he saw another car he would pull over to the left. Having driven through the Australian outback for years, he was comfortable with the protocol.

The countryside seemed to stretch out forever on that beautifully clear and breezy day, and the couple looked forward to a half hour or so of heading northwest through the Northern Territory until they arrived at the Western Australia border. Once there, they planned to pull off the road and have some lunch. The winter Sun shone from the north.

After a long stretch, the road finally began to curve gently and broadly. Off to the right, a pump of some sort rose above the desert. Carolyn gazed at the pump, casually noting the one artificial structure in miles of natural desert. Life didn't get any better than this: at the beginning of their twelfth season in Australia, they were, as Gene would have said, at the top of their game. Thirty-two comets race through the solar system bearing the name Shoemaker. Some two dozen craters on Earth are known to be impact structures, thanks to Gene Shoemaker. Five Surveyor missions and six manned landings on the Moon were accompanied by a lot of good field geology, thanks in part to the vision of Gene Shoemaker. Many thousands of people, including several hundred in Kingston and Wolfville less than three weeks earlier, knew why all this was important, thanks to Gene Shoemaker's ability to give a good lecture. Gene Shoemaker received some thirty major awards. As the Hilux rounded the curve, Gene could reflect on his life with dignity and pride.

And then, out of nowhere, a Land Rover materialized in front of them.

The Last Voyage

[A]nd, when he shall die,

Take him and cut him out in little stars,

And he will make the face of heaven so fine

That all the world will be in love with night,

And pay no worship to the garish sun.

—SHAKESPEARE, Romeo and Juliet, *1595.*

(These words are inscribed inside Lunar Prospector).

THREE YEARS LATER, it's still hard to believe that the greatest planetary scientist of this century, as Clark Chapman put it, is gone. Whenever I attend a meeting, or observe with Carolyn, I half expect to hear Gene's indomitable laugh. When Carolyn runs into a problem, she still finds herself asking, how would Gene handle this? Likewise, many scientists the world over, when they run into a problem, find themselves asking, how would Gene handle this?

Although Carolyn was seriously injured, she recovered sufficiently to return home to her "house of healing" in the middle of August 1997. On October 12, a large crowd gathered at the USGS in Flagstaff for a "celebration" of Gene's life in a rare Flagstaff October snowstorm. Since then Carolyn has been completing the papers Gene left behind; with the help of her daughter-in-law Paula Kempchinsky, Gene's Ruth Northcott lecture was transcribed, edited, and published in the *Journal of the Royal Astronomical Society of Canada* in December 1998.

38. Gene and Carolyn watch a tidal bore on the St. Croix River in Nova Scotia, on July 2, 1997, two weeks before Gene's death. Roy Bishop photo.

THE COMETOGRAPH GROWS AND PROSPERS

Using the Shoemaker-Levy Double Cometograph, Carolyn, Wendee, and I resumed observations, and we are now patrolling a portion of the sky for comets. We use the Flagstaff Observatory in the summer months, and our Jarnac Observatory in warmer Tucson the rest of the year. With each pair of films covering 78.5 square degrees of sky (an area covering almost half the constellation of Orion) we are getting good coverage of the sky. Well aware that the S-LDC was Gene's last "new start" project, we are pursuing it enthusiastically.

Gene knew that even though the S-LDC was intended primarily to keep Carolyn in the comet-finding business, if successful it will increase our understanding of the nature of comets, and of the chances that Earth will get struck by one in the near future. We see the project as a positive way of continuing the work that we shared for so long.

He Had a Dream

In 1948, when Gene was twenty years old, he decided he was going to the Moon. Forty-eight years later, while looking at the Moon through Echo, my own first telescope, he spoke about that dream:

> No, I'm not going to make it to the Moon. Just at the critical time when I should have been standing at the head of the line to go the Moon, my adrenal cortex stopped functioning. And I knew that would knock me out of the running medically. When you had that idea in your head for fifteen years, it doesn't go away.
>
> I was immensely pleased and proud of Jack [Schmitt], but of course I was wistful too. I couldn't help feeling that there, but for that failed adrenal gland, go I. For a long time after, I used to have dreams—thought I got there—got to the Moon, was there doing geology. I had to go do other things."[1]

The day after the accident, Carolyn Porco, a prominent planetary scientist and a Caltech student of Gene's, asked whether Gene's body would be cremated. I had remembered a conversation long ago, when he said that he preferred cremation. Porco had an idea. The *Lunar Prospector* spacecraft, she explained, was about to be sent for a year's orbit of the Moon, followed by an impact on the Moon. What could be more appropriate than sending a small portion of Gene's ashes to the Moon aboard the craft? On July 19, at Porco's request I described the idea to Carolyn, half a world away in an Alice Springs hospital, and I could feel her positive reaction. "Tell Carolyn," she said, "that the Shoemaker family would be thrilled if this tribute could happen."

Designed by scientists at Ames Research Center, *Lunar Prospector* was almost ready for launch to the Moon, and so Porco had to act quickly if the tribute was to be installed in the spacecraft before it was spin-balanced, an event scheduled for early fall. In late August, she drove from Tucson to Phoenix to oversee the fabrication of an inscription from *Romeo and Juliet*, then continued to Flagstaff to take part in a small family ceremony in which an ounce of ashes was selected from the remains of Gene's heart and soul. She then personally delivered the package to Ames Research Center, south of San Francisco, where mission manager Scott Hubbard coordinated the delicate task of placing the precious payload into the spacecraft before it was spin-balanced.

Lunar Prospector and *Athena 2* were finally slated for launch on January 5, 1998, at Pad 46 on Cape Canaveral. The family gathered at the Kennedy Space Center on Monday, January 5; Wendee and I were invited to join them. We waited at the stands that had been set up for us some three miles west of Pad 46. The sky was clear and the Moon was waiting, but a defective radar postponed the mission for twenty-four hours.

On the evening of January 6, all was ready except for a nearby thunderstorm; the launch area was at the southern edge of a horrible weather system that was paving northern areas, like Montreal, with sheets of ice; trees and power lines were falling like matchsticks. But the thunderstorm stayed to our west. Suddenly Wendee called out, "It's clearing!" We could see stars appearing to the east, and soon the first quarter Moon beckoned through the clouds. Launch control called out the final seconds, and then the countdown ended.

For a split second there was darkness and silence. Then the whole eastern horizon lit up with a bright orange glow. The dignified *Athena* rose from the launchpad and soared into the night, carrying with it the first space voyages of Gene Shoemaker and William Shakespeare, two thinkers who would now be forever bonded on the Moon.

By January 11, *Prospector* was in orbit a hundred kilometers above the Moon. In the next year, *Prospector* sent back strong evidence that there are quantities of ice in the Moon's polar regions, and it confirmed that the Moon has a small core, which adds support to the theory that the Moon was formed after a planetary body the size of Mars sideswiped the Earth.[2] "I read it in the newspaper that morning"—said Gene's granddaughter Stefani Salazar, "and faxed it to Grandma—see, even though Grandpa's dead, he's still busy doing what he loves!"

Early in 1999 *Lunar Prospector*'s orbit was lowered to a close twenty-four by thirty-seven kilometers (fifteen by twenty-three miles) above the Moon. Eventually it was decided to crash *Prospector* into the Moon's south pole, in the hope of liberating a bit of its meager layer of water so that Earth's satellite might briefly become a comet. Five years to the week after the last traces of Comet Shoemaker-Levy 9 joined with mighty Jupiter, *Prospector* dipped, like the *Ranger* spacecraft had done so long ago, and crashed into the lunar surface. On that night, Carolyn Porco joined Wendee and me outside with our telescopes. To the music of Antonín Dvořák's *New World Symphony*,[3] we watched the Moon as Gene finally came home, the man on the Moon, his dream fulfilled at last.

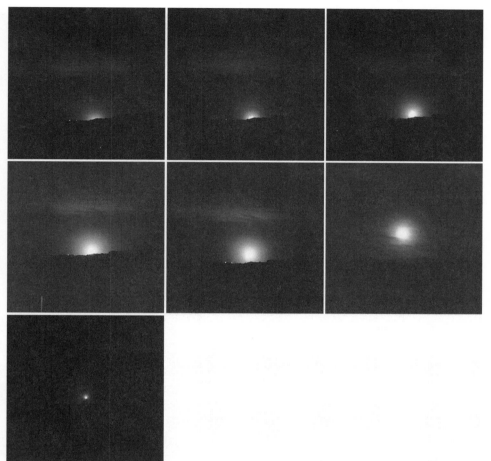

39. Up, up and away. This set of images shows the launch of *Lunar Prospector* to the Moon. Wendee Wallach-Levy photograph.

NOTES TO CHAPTER 1

1. Eugene. M. Shoemaker to Brian. G. Marsden (26 March 1993).
2. James V. Scotti to Brian G. Marsden (24 March 1993).

NOTES TO CHAPTER 2

1. George Shoemaker talked about Halley's Comet in a tape recording he made for Gene, circa 1959.
2. Maxine Heath—interview, 6 January 1998—provided material for this and other insights into Gene's childhood.
3. E. M. Shoemaker, interview with Fred Bortz, 30 June 1995, for *To the Young Scientist: Reflections on Doing and Living Science* (Franklin Watts, 1997).
4. Betty Lorain Shoemaker, contribution to Gene Shoemaker Album (1997).
5. Burt Shoemaker, contribution to Gene Shoemaker Album (1997).
6. Betty Lorain Shoemaker (1997).
7. E. M. Shoemaker, interview (22 October 1992).
8. Richard Spellman, interview, October 1998, and additional comments from other Caltech graduates.
9. Robert Sharp to Gene and Carolyn Shoemaker (2 August 1993).
10. William Muehlberger to Gene Shoemaker (5 October 1993); Muehlberger to Levy (2 September 1998).

NOTES TO CHAPTER 3

1. Ralph Baldwin, *The Face of the Moon* (Chicago: University of Chicago Press, 1949). See also Fred Whipple, *Earth, Moon, and Planets*, 3d ed. (Cambridge: Harvard University Press, 1968), 147.
2. G. Harry Stine, interview (10 July 1989).
3. C. W. Tombaugh, interview (1989).

Notes to Chapter 4

1. Grove Karl Gilbert, "The Origin of Hypotheses: Illustrated by the Discussion of a Topographic Problem," Presidential Address, Geological Society of Washington (March 1896). See also *Science*, NS 3 (1896), 1.
2. Ibid., 11.

Notes to Chapter 5

1. George L. Craik, *The Pursuit of Knowledge Under Difficulties* (London: Bell and Daldy, 1868), 2, 282.
2. Rollin Thomas Chamberlin, The Origin and Early Stages of the Earth, in Hortio H. Newman, ed., *The Nature of the World and of Man* (Chicago: University of Chicago Press, 1927). See also The Origin and History of the Earth, in Forrest Ray Moulton, *The World and Man As Science Sees Them* (New York: Literary Guild of America, 1937), and Stephen G. Brush, *Fruitful Encounters: The Origin of the Solar System and of the Moon from Chamberlin to Apollo* (Cambridge, England: Cambridge University Press, 1996).
3. Thomas Burnet, *Telluris Theoria Sacra*, or *The Sacred Theory of the Earth—Containing an Account of the Original of the Earth and of All the General Changes which it Hath Already Undergone or Is to Undergo till the Consummation of All Things* (1681; reproduced Carbondale: Southern Illinois University Press, 1965).
4. Claude C. Albritton Jr., *Catastrophic Episodes in Earth's History* (London: Chapman and Hall, 1989), 3.
5. Frank Dawson Adams, *The Birth and Development of the Geological Sciences* (1938; rpt. New York: Dover, 1954), 210.
6. J. G. Fitton, "On the Geological System of Werner," *Nicholson's Philosophical Journal*, vol. 36 (1813).
7. James Hutton, *Theory of the Earth with Proofs and Illustrations* (Edinburgh: Printed for Messrs. Cadell Jr., and Davies, 1795; rpt. New York: Stechert-Hafner Service Agency, 1972).
8. William Stokes and Sheldon Judson, *Introduction to Geology: Physical and Historical* (Englewood Cliffs, N.J.: Prentice-Hall, 1968), 146.
9. Adams, 250–55.
10. Johann Beringer, *Lithographiae Wirceburgensis* 1726, Melvin Jahn and Daniel Woolf, eds. (Berkeley: University of California Press, 1963).

11. Adams, 259–60.

12. Buffon, *Histoire naturelle des epoques de la nature. Histoire naturelle, générale et particulière. Supplément 5* (Paris: Imprimerie Royale, 1770, 1778), 99, 3.

13. Dorinda Outram, *George Cuvier: Vocation, Science, and Authority in Post-Revolutionary France* (Manchester: Manchester University Press, 19xx), 14–15.

14. Cuvier, *Discours*, 21.

15. George Cuvier, *Recherches sur les ossemens fossiles de quadrupedes* (1812; rpt. New York: Arno Press, 1980).

16. Cuvier, *Discours*, 20.

17. Charles Lyell, *Principles of Geology: Being an Attempt to Explain the Former Changes of the Earth's Surface by Reference to Causes Now in Operation* (London: J. Murray, 1830–33).

18. Derek Ager, *The New Catastrophism: The Importance of the Rare Event in Geological History* (Cambridge: Cambridge University Press, 1993), frontispiece.

19. Edgar Winston Spencer, *Basic Concepts of Historical Geology* (New York: Thomas Crowell, 1962, 1971), 51.

20. Alfred Wegener, *The Origins of Continents and Oceans* (1929; rpt. New York, Dover, 1966).

21. Walter Alvarez: *T. Rex and the Crater of Doom* (Princeton: Princeton University Press, 1997), 56.

22. Reijer Hooykaas, *Catastrophism in Geology: Its Scientific Character in Relation to Actualism and Uniformitarianism* (Amsterdam: North Holland Publishing Company, 1970), 50.

23. See André Goddu, *The Physics of William of Occam* (Leiden: E. J. Brill, 1984.)

24. Thomas Kuhn, *The Structure of Scientific Revolutions* (Chicago: University of Chicago Press, 1962), 172.

25. W. A. Berggren and John A. Van Couvering, eds., *Catastrophes and Earth's History: The New Uniformitarianism* (Princeton: Princeton University Press, 1984), 16.

26. Stephen Jay Gould, *The Panda's Thumb* (New York: Norton, 1980), 188.

27. George Stevens, personal communication 1995.

28. George Wetherill and Eugene Shoemaker, Collision of Astronomically Observable Bodies with the Earth, in Leon T. Silver and Peter H. Schultz, eds., *Geological Implications of the Impacts of Large Asteroids and Comets on the Earth* (Special Paper 190: The Geological Society of America, 1982), 1.

NOTES TO CHAPTER 6

1. Diana Brueton, *The Moon: Myth, Magic, and Fact* (New York: Barnes & Noble, 1991), 80, 139.

2. Grove Karl Gilbert, *Philosophical Society of Washington Bulletin* 12, No. 242 (1893).

3. Baldwin (1949).

4. J. Russell Hind, *The Comet of 1556: Being Popular Replies to Everyday Questions, Referring to Its Anticipated Reappearance, With Some Observations on the Apprehension of Danger from Comets* (London: John W. Parker and Son, 1857), 45.

5. Immanuel Velikovsky, *Worlds in Collision* (New York: Macmillan, 1950).

6. Trevor Palmer, *Catastrophism, Neocatastrophism, and Evolution* (Nottingham: The Society for Interdisciplinary Studies, 1994).

7. Steven M. Stanley, *Macroevolution: Patterns and Process* (San Francisco: Freeman, 1979), 35.

8. Max Walker de Laubenfels: "Dinosaur Extinction—One More Hypothesis," *Journal of Paleontology* 30 (1956), 207–18.

9. "Neokatastrophismus?" *Deutsche Geologisch Gesellschaft Zeitschrift*, 114 (1963), 430–645.

10. Harold C. Urey, "Cometary Collisions and Geologic Periods," *Nature* 242 (1973), 32–33.

11. D. A. Russell, "The Environments of Canadian Dinosaurs," *Canadian Geographical Journal* 87 (1973), 4–11.

12. H. J. Hsü, "Terrestrial Catastrophe Caused by Cometary Impact at the End of the Cretaceous," *Nature* 285 (1980), 201–203.

13. Luis W. Alvarez, Walter Alvarez, F. Asaro, and H. V. Michel, "Extraterrestrial Cause for the Cretaceous-Tertiary Extinction, *Science* 208 (1980) 1095–108. Walter Alvarez's book *T. Rex and the Crater of Doom* (Princeton: Princeton University Press, 1997) is an accurate and well-written account of the complex story that led to our understanding of what happened in the last day of the Mesozoic era.

14. Charles B. Officer and C. L. Drake, "Terminal Cretaceous Environment Events," *Science* 227 (1985), 1161–67.

15. See Adriana Ocampo, Kevin Pope, Michael Rampino, Alfred Fischer, and David Kring Jr., in *The Planetary Report* 16, 4 (July/August 1996).

16. David A. Kring, A. R. Hildebrand, and W. V. Boynton, "The Petrology of an Andesitic Melt Rock and a Polymict Breccia from the Interior of the Chicxulub Structure, Yucatán, Mexico," *Lunar and Planetary Science* 22 (1991), 755–56. See also D. A. Kring and W. V. Boynton, "The Petrogenesis of an Augite-Bearing Melt Rock in the Chicxulub Structure and Its Relationship to K/T Impact Spherules in Haiti," *Nature* 358 (1992),141–44.

17. Stephen K. Donovan, ed., *Mass extinctions: Processes and Evidence* (London: Belhaven Press, 1989), xii.

18. See George R. McGhee, Jr. *The Late Devonian Mass Extinction: The Frasnian-Famennian Crisis* (New York: Columbia University Press, 1996), 6.

19. W. Desmond Maxwell, "The End Permian Mass Extinction," in Stephen K. Donovan, ed. *Mass Extinctions: Processes and Evidence* (London: Belhaven Press, 1989), 152–73.

20. Roger Osborne and Donald Tarling, eds. *The Historical Atlas of the Earth: A Visual Exploration of the Earth's Physical Past* (New York: Henry Holt, 1996), 87.

21. Michael R. Rampini and Bruce M. Haggerty, "Extraterrestrial Impacts and Mass Extinctions," *Hazards*, 841.

22. *Nature*, 12 March 1998.

23. Eugene M. Shoemaker, Ruth Northcott Lecture, Royal Astronomical Society of Canada, Kingston, Ontario (30 June 1997).

24. J. Kelly Beatty, "Special Report: Of Comets and Cataclysms," *Sky and Telescope* 67 (1984), 406.

NOTES TO CHAPTER 7

1. C. Shoemaker and Phred Salazar, interview 24 April 1999.

2. Gilbert, *Science*, NS 3 (1896), 11.

3. E. M. Shoemaker, R. J. Hackman, and R. E. Eggleton, "Interplanetary Correlation of Geologic Time," in *Advances in the Astronautical Sciences*, vol. 8 (New York: Plenum, 1962), 70–89.

4. Richard F Grieve, "Meteorite Impact Craters of North America," *Observer's Handbook 1998* (Toronto: The Royal Astronomical Society of Canada, 1998), 186–88.

5. E. M. Shoemaker, personal retrospective about his father, written in the early 1990s.

NOTES TO CHAPTER 8

1. Loring Coes Jr., "A New Dense Crystalline Silica Mineral," *Science* 118 (1953) 131–33.
2. Don E. Wilhelms, *To a Rocky Moon: A Geologist's History of Lunar Exploration* (Tucson: University of Arizona Press, 1993), 45.
3. Edward Chao, E. M. Shoemaker, and Beth M. Madsden, "First Natural Occurrence of Coesite from Meteor Crater, Arizona, *Science* 132 (3421) (1960), 220–22.
4. E. M. Shoemaker, Ballistics of the Copernican Ray System, Lunar and Planetary Exploration Colloquium, Proceedings, vol. 2, no. 2, 7–21.
5. C. S. Shoemaker, Christmas letter (December 1960).
6. Some trivia: The word *program* referred to a program as administered at NASA headquarters, while the word *project* referred to the work conducted at a field center.
7. Stephen Dwornik, interview (27 July 1999).
8. Linda Salazar, interview (24 April 1999).
9. *Asteroids: Deadly Impact.* Eitan Weinreich, producer; National Geographic Society; NBC (February 1997).

NOTES TO CHAPTER 9

1. The *Journal of the Association of Lunar and Planetary Observers* reported on its Lunar Meteor Search over several issues. The program was followed by observers all over North America, as evidenced by the Montreal Centre of the Royal Astronomical Society's *Skyward* (January 1964), 3, and (November 1964), 2.
2. Wilhelms, 97.

NOTES TO CHAPTER 10

1. "Moon" episode of *The Planets*, BBC/A&E television documentary (16 October 1999).
2. Ibid.
3. *Daily Express* (5 February 1966).
4. American Geological Institute, *Dictionary of Geological Terms* (New York: Dolphin, 1952), 419.
5. E. M. Shoemaker, interview (27 July, 1993).

6. Justin Rennilson, interview (8 November 1998). See E. M. Shoemaker, L. D. Jaffe, et al., "Surveyor 1: Preliminary Results," *Science* 152, no. 3730 (1966) 1737–50; also published in *Surveyor 1: A Preliminary Report*, National Aeronautics and Space Administration, Special Publication 126, Washington, D.C. (1966).

7. Wilhelms, 145.

8. Lee Silver, interview (30 August 1999).

9. Steve Dwornik, interview (27 July, 1999).

10. Rennilson, "Colorimetric Measurements of the Solar Eclipse and Earth from *Surveyor 3*," *JPL Technical Report 32–1443* (Pasadena: NASA/JPL, 1967).

11. C. S. Shoemaker, interview (11 November 1998).

12. Wilhelms, 154.

13. Henry Holt, interview (10 May 1999).

14. Wilhelms, 146–47.

Notes to Chapter 11

1. Hal G. Stephens and E. M. Shoemaker, *In the Footsteps of John Wesley Powell: An Album of Comparative Photographs of the Green and Colorado Rivers, 1871–72 and 1968* (New York: Johnson Books, 1987).

2. Gordon Swann, interview (8 June, 1999).

3. Wilhelms, 192–93.

4. At least they did on CBS's coverage that night.

5. *Time* (3 January 1969).

6. Swann, 1999.

7. C. S. Shoemaker, annual letter (December 1969).

8. Wilhelms, 204.

9. Wilhelms, 205.

Notes to Chapter 12

1. Swann (1999).

2. E. M. Shoemaker, "Space: Where Now, and Why?" *Engineering and Science* (October 1969).

3. Dwornik (1999).

4. The relevant papers:

E. M. Shoemaker, N. G. Bailey, R. M. Batson, D. H. Dahlem, T. H. Foss, M. J. Grolier, E. N. Goddard, M. H. Hait, H. E. Holt, K. B. Larson,

J. J. Rennilson, G. G. Schaber, D. L. Schleicher, H. H. Schmitt, R. L. Sutton, G. A. Swann, A. C. Waters, and M. H. West, "Geologic Setting of the Lunar Samples Returned by the *Apollo 11* Mission," Apollo 11 Preliminary Science Report: National Aeronautics and Space Administration Special Publication 214 (1969), 41–83.

E. M. Shoemaker, M. H. Hait, G. A. Swann, D. L. Schleicher, D. H. Dahlem, G. G. Schaber, R. L. Sutton, "Lunar Regolith at Tranquility Base," *Science* 167 (1970), 452–55.

E. M. Shoemaker, R. M. Batson, A. L. Bean, C. Conrad, D. H. Dahlem, E. N. Goddard, M. H. Hait, K. B. Larson, G. G. Schaber, D. L. Schleicher, R. L. Sutton, G. A. Swann, and A. C. Waters, 1970, "Preliminary Examination of Lunar Samples from *Apollo 12*," *Science* 167 (1970), 1325–39.

E. M. Shoemaker, M. H. Hait, G. A. Swann, D. L. Schleicher, G. G. Schaber, R. L. Sutton, D. H. Dahlem, E. N. Goddard, and A. C. Waters, "Origin of the Lunar Regolith at Tranquility Base," Apollo 11 Lunar Science Conference, Proceedings, vol. 3, Supplement 1 (1970), 2399–412.

5. Wilhelms, 326.

6. Richard Brautigan, *Trout Fishing in America* (Boston: Houghton Mifflin, 1968).

7. Schmitt (1999).

8. E. M. Shoemaker, interview (17 February 1993).

9. Wilhelms, 57.

10. Wilhelms, 313–14.

11. Wilhems, 107.

12. Wilhelms, 347.

13. E. M. Shoemaker, interview (27 July 1993).

14. Dwornik (1999).

15. E. M. Shoemaker, interview (27 July 1993).

NOTES TO CHAPTER 13

1. C. S. Shoemaker, interview (16 October 1999).

2. Patrick Shoemaker, interview (6 January 1998).

3. Linda Salazar, interview (24 April 1999).

4. Ibid.

5. Paula Kempchinsky, interview (6 January 1998).

6. Phred Salazar, interview (24 April 1999).

7. Ronald Greeley, interview (27 July 1999).

8. Silver (1999).

9. Job 38: 24–25, 31.

10. Larry Lebofsky, interview (12 February 1999).

11. Carolyn Porco, interview (15 February 1999).

12. Ibid.

13. Donald Brownlee, personal communication (15 February 2000).

14. Joseph Kirschvink, interview (29 September 1999).

15. Susan Kieffer, interview (10 September, 1999).

16. Ibid.

17. Shakespeare, *Julius Caesar* act 4, scene 3, line 97. *The Riverside Shakespeare* (Boston: Houghton Mifflin, 1974), p. 1126.

18. Robert Sharp, interview (October 1998).

19. Silver (1999).

20. Kieffer (1999).

21. Margaret Marsh, interview (1 August 1999).

22. Sharp (1998).

23. Shakespeare, *All's Well That Ends Well* act 4, scene 4, line 35. *The Riverside Shakespeare* (Boston: Houghton Mifflin, 1974), p. 535.

24. C. S. Shoemaker, interview (21 January, 1999).

25. Larry Lebofsky (1999).

Notes to Chapter 14

1. Silver (1999).

2. E. M. Shoemaker and R. J. Hackman, "Stratigraphic Basis for a Lunar time Scale," Kopal, Zdenek, and Mikhailov, Z. K., eds., *The Moon*, Symposium no. 14 of the International Astronomical Union: London, Academic Press (1962), 289–300.

3. Wilhelms, 220.

4. E. M. Shoemaker, D. P. Elston, and C. E. Helsley "Depositional History of the Moenkopi Formation in Light of Its Magnetostratigraphy" (abstract) *Geological Society of America*, Abstracts with Programs, vol. 5, no. 7 (1973), 807–808.

5. Walter Alvarez's book *T Rex and the Crater of Doom* (Princeton: Princeton University Press, 1997, 34–40) contains a well-written account of the discovery of magnetic pole reversals in the Scaglia rossa limestone and how it related to the complex story that led to our understanding of what happened in the last day of the Mesozoic era.

6. J. N. Kellogg and E. M. Shoemaker, "Age Determination of Volcanic Rocks by Spatial Frequency of Lightning Strikes in the San Francisco Vol-

canic Field, Arizona," Transactions of the American Geophysical Union, *Eos* 58 (6) (1977), 376–77.

7. Kirschvink (1999).

8. D. Elston, interview (8 June 1999).

9. Ibid.

NOTES TO CHAPTER 15

1. Antoine de Saint Exupéry, *The Little Prince* (San Diego: Harcourt Brace Jovanovich, 1943, 1971).

2. E. M. Shoemaker, Robert J. Hackman, and Richard E. Eggleton, "Interplanetary Correlation of Geologic Time," *Advances in the Astronautical Sciences*, 8 (1963), 70–89.

3. Eleanor F. Helin, personal communication (7 July 1999).

4. E. M. Shoemaker, interview (July 1992).

5. Berton C. Willard, *Russell W. Porter: Arctic Explorer, Artist, Telescope Maker* (Freeport, Maine: Bond Wheelwright, 1976), 145.

6. Albert G. Ingalls, "The Heavens Declare the Glory of God," *Scientific American* (November 1925), 293–95.

7. Willard, 187.

8. Gene Shoemaker, interview (11 February 1993).

9. Helin (1999).

10. Helin, interview (July 1993).

11. Stuart J. Weidenschilling, Clark R. Chapman, Donald R. Davis, Richard Greenberg, David H. Levy, and Sheila Vail, "Photometric Geodesy of Main-Belt Asteroids, I: Light Curves of Twenty-six Large, Rapid Rotators," *Icarus* 70 (1987), 191–245. See also, Jack D. Drummond, S. J. Weidenschilling, C. R. Chapman, and D. R. Davis, "Photometric Geodesy of Main-Belt Asteroids, ii: An Analysis of Light Curves for Poles, Periods, and Shapes," *Icarus* 76 (1988), 19–77; and S. J. Weidenschilling, C. R. Chapman, D. R. Davis, R. Greenberg, D. H. Levy, Richard P. Binzel, S. Vail, Michael Magee, and Dominique Spaute, "Photometric Geodesy of Main-Belt Asteroids, iii: Additional Light Curves," *Icarus* 86 (1990), 402–47.

12. Eleanor. F. Helin and E. M. Shoemaker, "The Palomar Planet-Crossing Asteroid Survey, 1973–78," *Icarus* 40 (1979), 321–28.

13. David H. Levy, *Clyde Tombaugh: Discoverer of Planet Pluto* (Tucson: University of Arizona Press, 1991), 2–7, 49–50.

14. Robert Hash, McBain Instruments, Inc., to Eleanor Helin (29 June 1978).

15. Helin (1999).

16. Hash to Helin (24 December 1980).

17. Lutz D. Schmadel, *Dictionary of Minor Planet Names* (Berlin, Springer-Verlag, 1992), 593; cf. *Minor Plant Circular* 19338.

18. Helin (1993).

19. *Time* (8 May 1985).

20. E. M. Shoemaker, interview with Fred Bortz (30 June 1995).

NOTES TO CHAPTER 16

1. E. M. Shoemaker, interview (17 February 1993).

2. E. M. Shoemaker, interview (1993).

3. Bonnie P. Buratti, "Outer Planet Icy Satellites," in Paul R. Weissman, Lucy-Ann McFadden, and Torrence V. Johnson, *Encyclopedia of the Solar System* (San Diego: Academic Press, 1998), 435–55.

4. David Morrison and M. Shapley Matthews, *Satellites of Jupiter* (Tucson: University of Arizona Press, 1981).

5. William B. McKinnon, "Midsize Icy Satellites" in *The New Solar System*, J. Kelly Beatty, Carolyn Collins Petersen, and Andrew Chaikin, eds. (Cambridge, Mass., and Cambridge, England: Sky Publishing Corporation and Cambridge University Press), 308.

6. Shakespeare, *The Tempest*, in *The Riverside Shakespeare* (Boston: Houghton Mifflin, 1974) p. 1634. Act 5, scene 1, lines 181–84.

NOTES TO CHAPTER 17

1. C. S. Shoemaker, interview (30 November 1992).

2. J. K. Beatty, *Sky and Telescope* 67 (1984), 406.

3. Clark Chapman, personal communication (1 June 1999).

NOTES TO CHAPTER 18

1. C. S. Shoemaker, interview (17 February 1999).

2. Australian crater ages are derived from Richard A. F. Grieve and Eugene M. Shoemaker, "The Record of Past Impacts on Earth," *Hazards Due to Comets and Asteroids*, Tom Gehrels, ed. (Tucson: University of Arizona Press, 1994), 417–62.

3. C. S. Shoemaker, Australian journal (August 1984).

4. Candace Kohl, interview (3 February 1998).

5. C. S. Shoemaker, interview (17 March 1999).

6. Carolyn Shoemaker (August 1984).

7. Kohl (1998).

8. David Taylor, *Three Minutes t o Impact*, Discovery channel documentary (aired 9 February 1997).

9. Carolyn Shoemaker, interview (19 March 1999).

10. Sean Woodard, interview (26 April 1999).

11. C. S. Shoemaker, "The Ups and Downs of Planetary Science, *Annual Review of Earth and Planetary Science*, in press.

NOTES TO CHAPTER 19

1. C. S. Shoemaker, interview with Fred Bortz (30 June 1995).

2. Brian G. Marsden, IAU *Circular 5507*, Central Bureau for Astronomical Telegrams (26 March 1993).

3. James V. Scotti, interview (2 March 1994).

4. H. J. Melosh and P. Schenk, "Split Comets and the Origin of Crater Chains on Ganymede," *Nature* 265 (1993), 731–33.

5. H. J. Melosh and E. A. Whitaker, "Split Comets and Crater Chains on the Moon, *Nature* 369 (1994) 713–14.

6. Marsden, *Circular 5744*.

7. Ibid. personal communication (January 1989).

8. Briefly, these orbital elements are *T* for comet's closest approach to the Sun (which in this case matches Jupiter's closest approach to the Sun,) *e* for how elliptical the orbit is, *q* for the comet's distance from the Sun on the epoch date, *Incl.* for how inclined the orbit is to the plane of the ecliptic, and *P* for the comet's period of revolution about the Sun. *Peri.* is an abbreviation for argument of perihelion, *Node* means longitude of the orbit's ascending node, *a* is the semimajor axis in astronomical units, where one AU is the distance between Earth and Sun, and *n* is the rate of change of the mean anomaly per day—"n" is a means of defining where the comet is in its orit.

9. Marsden, *Circular 5800* (22 May 1993).

10. Ibid., *Circular 5801*, 22 May 1993.

11. Paul W. Chodas and Donald K. Yeomans, "The Orbital Motion and Impact Circumstances of Comet Shoemaker-Levy 9," in *The Collision of Comet Shoemaker-Levy 9 and Jupiter*, Keith Knoll, Harold

Weaver, and Paul Feldman, eds. (Cambridge, England: Cambridge University Press, 1996), 1–30.

12. H. J. Melosh, "Report on the 'Comet Pre-Crash Bash' " (20 September 1993).

13. Donald Gault to Larry Soderblom (14 August 1993).

14. Bevan French, *The Man Passing By on His Way to the Moon*, sung October 1993 at Gene's retirement from the U.S. Geological Survey, Flagstaff, Arizona.

15. Naomi and Jerry Wasserburg to Gene and Carolyn Shoemaker (25 October 1993).

NOTES TO CHAPTER 20

1. E. M. Shoemaker, Boulder, Colo. (18 October 1998).

2. Jody Swann, interview (8 June, 1999).

3. Carolyn Shoemaker, interview (6 February 2000).

4. Clark Chapman, personal communication (1 June 1999).

5. Carl Sagan, Boulder (18 October 1993).

6. Harold Weaver, personal communication (23 February 1995).

7. Paul Weissman, "Comet Shoemaker-Levy 9: The Big Fizzle is coming" *Nature* 370 (1994), 94–95.

8. Weaver (23 February 1995).

9. Alan Harris, interview (26 April 1999).

10. Neil Armstrong, White House ceremony (20 July 1994).

11. Clark R. Chapman, S-L9 exploder (an e-mail notification system for scientists studying Comet S-L9 and Jupiter), University of Maryland (18 July 1994).

12. Peter Jennings, *Person of the Week*, ABC Nightly News (July 22, 1994).

NOTES TO CHAPTER 21

1. Gordon Swann (1999).

2. Eric Asphaug and Willy Benz, "Size, Density, and Structure of Comet Shoemaker-Levy 9 Inferred from the Physics of Tidal Breakup," *Icarus* 121 (1995), 225–48.

3. Mordecai-Mark Mac Low, "Entry and Fireball Models vs. Observations: What Have We Learned?" in K. S. Noll, H. A. Weaver, P. D. Feld-

man, eds. *The Collision of Comet Shoemaker-Levy 9 and Jupiter* (1996), 157–82.

4. Zdenek Sekanina, "Tidal Breakup of the Nucleus of Comet Shoemaker-Levy 9," in ibid., 55–80.

5. Zdenek Sekanina, Paul W. Chodas, and Donald K. Yeomans, "Secondary Fragmentation of Comet Shoemaker-Levy 9 and the Ramifications for the Progenitor's Breakup in July 1992," *Planetary and Space Science* 46 (1998), 21–45.

6. Chodas and Yeomans, 1–30.

7. Juliane I. Moses, "Long-Term Photochemical Evolution of the Impact Sites," presented at the International Conference on the SL9-Jupiter Collision, Meudon, France (July 1996).

8. P. D. Wilson and Carl Sagan, "Nature and Source of Organic Matter in the Shoemaker-Levy 9 Jovian Impact Blemishes," *Icarus* 129 (1997), 207–16.

9. Eugene M. Shoemaker, "Summary of Conference," presented at Collision of Comet Shoemaker-Levy 9 and Jupiter Conference, Baltimore, Md. (May 1995).

10. M. S. Roulston and Thomas J. Ahrens, "Impact Mechanics and Frequency of S-L9–type events on Jupiter," *Icarus* 126 (1997), 138–47.

11. Eugene M. Shoemaker "Summary of Conference," Baltimore, Md., (May 1995).

12. Jeffrey C. Wynn and Eugene M. Shoemaker, "Secrets of the Wabar Craters," *Sky and Telescope* 94 (November 1997), 44–48.

13. Carolyn Shoemaker, interview (19 March 1999).

Notes to Chapter 22

1. Carolyn Shoemaker, interview (19 March 1999).

2. Brian G. Marsden, *Minor Planet Circular* 27330 (1 June 1996).

3. David Taylor, Eugene M. Shoemaker, and David H. Levy, "Origins: A Shoemaker-Levy Tour of the Universe." Proposal for television series (1997).

4. Gene had recently read and enjoyed James Shreeve's *The Neandertal Enigma: Solving the Mystery of Modern Human Origins* (New York: William Morrow, 1995).

5. Christy Abanto, interview (25 April, 1999).

6. Carolyn Porco, interview (16 February 1999).

7. Harris (1999).

8. E. M. Shoemaker, "Impact Cratering through Geologic Time," *The Journal of the Royal Astronomical Society of Canada* 92 (1998) 297–309.

NOTES TO EPILOGUE

Shakespeare, *Romeo and Juliet*, *The Riverside Shakespeare* (Boston: Houghton Mifflin, 1974) III, ii, 20–25, p. 1077. Except for the Fourth Quarto edition of this play, the original texts have the second line of the quote saying "Give me my Romeo, and when I shall die," a statement that significantly alters the meaning of the text. The second quarto appeared in 1599 and is considered the authoritative text, but it does have some problems, including this one, that the fourth quarto attempted to correct. The text that was sent as Shakespeare's first voyage into space and to the Moon follows:

> And, when he shall die,
> Take him and cut him out in little stars,
> And he will make the face of heaven so fine
> That all the world will be in love with night,
> And pay no worship to the garish sun.

1. Eitan Weinreich, producer, *Asteroids: Deadly Impact*, National Geographic Television, aired on NBC February 1997.
2. Ames Research Center, NASA release 99–43 (16 March 1999).
3. Antonín Dvořák, *From the New World*, Symphony in E Minor, op. 95 (1893).

SELECTED BIBLIOGRAPHY

Eugene M. Shoemaker

Gene Shoemaker's prodigious scientific output includes 195 publications and more than 200 abstracts. For this bibliography, I have included a selection of about one hundred of the papers that he wrote or contributed to, including three papers written by Carolyn alone. The list attempts to show the variety of subjects that captured his interest. I am indebted to Gene and Carolyn for their preparation of material for this bibliography.

1953

Thirty selected papers—an annotated bibliography of the Colorado Plateau. U.S. Geological Survey, 6 p.

1954

"Structural Features of Southeastern Utah and Adjacent Parts of Colorado, New Mexico, and Arizona." Utah Geological Society. *Guidebook to the Geology of Utah* no. 9, 48–69.

1955

"Preliminary Geologic Map of the Juanita Arch Quadrangle, Colorado." U.S. Geological Survey Mineral Investigation Field Studies Map MF-28.
"Geology of the Juanita Arch Quadrangle, Colorado." U.S. Geological Survey Quadrangle Map GQ 81.
"Preliminary Map of the Rock Creek Quadrangle, Colorado." U.S. Geological Survey Mineral Investigation Field Studies Map MF-23.

1956

"Geology of the Rock Creek Quadrangle, Colorado." U.S. Geological Survey Quadrangle Map GQ 83.
"Occurrence of Uranium in Diatremes on the Navajo and Hopi Reservations, Arizona, New Mexico, and Utah." United Nations, Geology of Uranium and

Thorium: International Conference on Peaceful Uses of Atomic Energy. Geneva. August 1956 Proceedings. V. 6, 412–17. Slightly revised, in Page, L. R., "Contributions to the Geology of Uranium and Thorium . . ." U.S. Geological Survey Professional Paper 300, 179–85.

"Precambrian Rocks of the North-Central Colorado Plateau." Intermountain Association of Petroleum Geologists Field Conference. Seventh Annual Field Conference, 54–59.

"Structural Features of the Central Colorado Plateau and Their Relation to Uranium Deposits." In Page, L. R., "Contributions to the Geology of Uranium and Thorium " U.S. Geological Survey Professional Paper 300, 155–70.

1958

with Case, J. E., and Elston, D. P. "Salt Anticlines of the Paradox Basin." *Intermountain Association of Petroleum Geologists, Guidebook*. Ninth, Annual Field Conference, 39–59.

1959

with Newman, W. L. "Moenkopi Formation in the Salt Anticline Region, Colorado and Utah." *American Association of Petroleum Geologists Bulletin*. Vol. 43, no. 8, 1835–51.

1960

with Chao, E. C. T., and Madsen, B. M. "First Natural Occurrence of Coesite from Meteor Crater, Arizona." *Science*. Vol. 132, no. 3421, 220–22.

with Miesch, A. T., Newman, W. L., and Finch, W. I. "Chemical Composition As a Guide to the Size of Sandstone-Type Uranium Deposits in the Morrison Formation on the Colorado Plateau." *U.S. Geological Survey Bulletin* 1112-B, 17–61.

"Ballistics of the Copernican Ray System." Lunar and Planetary Exploration Colloquium, Proceedings. Vol. 2, no. 2, 7–21.

"Brecciation and Mixing of Rock by Strong Shock." Article 192. U.S. Geological Survey Professional Paper 400-B, B423–25.

"Penetration Mechanics of High Velocity Meteorites, Illustrated by Meteor Crater, Arizona." Twenty-first International Geological Congress. Copenhagen. Report, pt. 18, 418–34.

1961

with Eggleton, R. E. "Breccia at Sierra Madera, Texas." Article 342. U.S. Geological Survey Professional Paper 424-D, D151–53.

"Ballistics and Throw-Out Calculations for the Lunar Crater Copernicus." Proceedings of the Geophysical Laboratory/Lawrence Radiation Laboratory Cratering Symposium. Washington, D.C. Pt. 2. University of California, Livermore. Report UCRL-6438, Paper Q, p. 31 (Report prepared for U.S. Atomic Energy Commission.)

with Chao, E. C. T. "New Evidence for the Impact Origin of the Ries Basin, Bavaria, Germany." *Journal of Geophysical Research*. Vol. 66, no. 10, 3371–78.

1962

with Roach, C. H., and Byers, F. M., Jr. "Diatremes and Uranium Deposits in the Hopi Buttes, Arizona." *Petrologic Studies, A Volume to Honor A. F. Buddington*. Geological Society of America, 327–55.

with Hackman, R. J. Stratigraphic Basis for a Lunar Time Scale. In Kopal, Zdenek, and Mikhailov, Z. K., eds. *The Moon*—Symposium no. 14 of the International Astronomical Union. London: Academic Press, 289–300.

1963

"Impact Mechanics at Meteor Crater, Arizona." In Middlehurst, B., and Kuiper, G. P., eds. *The Moon, Meteorites, and Comets*—The Solar System. Vol. 4: Chicago: University of Chicago Press, 301–36.

with Hackman, R. J., and Eggleton, R. E. "Interplanetary Correlation of Geologic Time. In *Advances in the Astronautical Sciences*. Vol. 8: New York: Plenum Press, 70–89.

1964

"The Moon Close Up." *National Geographic*. Vol. 126, no. 5, 690–707.

"The Geology of the Moon." *Scientific American*. Vol. 211, no. 6, 38–47.

1965

"Preliminary Analysis of the Fine Structure of the Lunar Surface in Mare Cognitum, in *Ranger 7*." Pt. 2. Experimenters' Analyses and Interpretations. Jet Propulsion Laboratory. California Institute of Technology. Technical Report No. 32–700, 75–134.

1966

"When the Irresistible Force Meets the Immovable Object." *Engineering and Science*. Vol. 29, no. 5, 11–15.

"Interpretation of the Small Craters of the Moon's Surface Revealed by *Ranger* 7. Transactions of the International Astronomical Union. General Assembly. Proceedings, 12. Hamburg, Germany. Vol. XIIB, 662–72.

with Batson, R. M., and Larson, K. B. "An appreciation of the *Luna 9* Pictures." *Astronautics and Aeronautics*. May 1966, 40–50.

with Jaffe, L. D. "*Surveyor I*: Preliminary Results." *Science* 152. No. 3730, 1737–50; also published in *Surveyor 1: A preliminary report*. National Aeronautics and Space Administration. Special Publication 126. Washington, D.C.

1967

with Batson, R. M., Holt, H. E., Morris, E. C., Rennilson, J. J., and Whitaker, E. A. "*Surveyor 5*: Television pictures." *Science*. Vol. 158, no. 3801, 642–52.

1968

with Gault, D. E.; Adams, J. B.; Collins, R. J.; Kuiper, G. P.; Masursky, Harold; O'Keefe, J. A.; Phinney, R. A. "Lunar Theory and Processes." *Journal of Geophysical Research*. Vol. 73, no. 12, 4115–31.

with Batson, R. M., Holt, H. E., Morris, E. C., Rennilson, J. J., and Whitaker, E. A. "Television Observations from *Surveyor 3*." *Journal of Geophysical Research*. Vol. 73, no. 12, 3989–4043.

with Morris, E. C., Batson, R. M., Holt, H. E., Larson, K. B., Montgomery, D. R., Rennilson, J. J., and Whitaker, E.A. "Television Observations from *Surveyor*. In *Surveyor Project Final Report*, Part 2: Science Results: Jet Propulsion Laboratory, California Institute of Technology. Technical Report No. 32–1265, 21–136; also published in National Aeronautics and Space Administration Special Publication 184, 351–67.

1969

"The Lunar Regolith." In Randall, C. A. Jr., ed. Extraterrestrial Matter. Conference at Argonne National Laboratory. Proceedings. March 7–8. Northern Illinois University Press, 93–136.

"Space—Where, Now, and Why?" *Engineering and Science*. Vol. 33, no. l, 9–12.

with Jaffe, L. D. (chairman); Alley, C. O.; Batterson, S. A.; Christensen, E. M.; Dwornik, Gault; D. E., Lucas; J. W.; Muhleman, D. O.; Norton, R. H.; Scott, R. F.; Steinbacher, R. H.; Sutton, G. H., and Turkevich, A. L. Principal Scientific Results from the Surveyor Program." In *Surveyor Program Results*. National Aeronautics and Space Administration Special Publication 184, 13–17.

with Phinney, R. A.; O'Keefe, J. A.; Adams, J. B.; Gault, D. E.; Kuiper, G.P.; Masursky, Harold; Collins, R. J. "Implications of the *Surveyor 7* Results." *Journal of Geophysical Research*. Vol. 74, no. 25, 6053–80.

with Bailey, N. G., Batson, R. M., Dahlem, D. H., Foss, T. H., Grolier, M. J., Goddard, E. N., Hait, M. H., Holt, H. E., Larson, K. B., Rennilson, J. J., Schaber, G. G., Schleicher, D. L., Schmitt, H. H., Sutton, R. L., Swann, G. A., Waters, A. C., and West, M. H. "Geologic Setting of the Lunar Samples Returned by the *Apollo 11* Mission. In Apollo 11 Preliminary Science Report: National Aeronautics and Space Administration Special Publication 214, 41–83.

with other members of the Lunar Sample Preliminary Examination Team. "Preliminary Examination of Lunar Samples from *Apollo 11*." *Science*. Vol. 165, no. 3899, 1211–27.

1970

with Batson, R. M., Bean, A. L., Conrad, C., Dahlem, D. H., Goddard, E. N., Hait, M. H., Larson, K. B., Schaber, G. G., Schleicher, D. L., Sutton, R. L., Swann, G. A., and Waters, A. C. "Preliminary Geologic Investigation of the *Apollo 12* Landing Site, in Apollo 12, Preliminary Science Report. National Aeronautics and Space Administration Special Publication 235.

with Hait, M. H., Swann, G. A., Schleicher, D. L., Dahlem, D. H., Schaber, G. G., and Sutton, R. L. "Lunar Regolith at Tranquility Base." *Science*. Vol. 167, no. 3918, 452–55.

with Batson, R. M., Bean, A. L., Conrad, C., Dahlem, D. H., Goddard, E. N., Hait, M. H., Larson, K. B., Schaber, G. G., Schleicher, D. L., Sutton, R. L., Swann, G. A., and Waters, A. C. "Preliminary Examination of Lunar Samples from *Apollo 12*." *Science*. Vol. 167, no. 3923, 1325–39.

with Hait, M. H., Swann, G. A., Schleicher, D. L., Schaber, G. G., Sutton, R. L., Dahlem, D. H., Goddard, E. N., and Waters, A. C. "Origin of the Lunar Regolith at Tranquility Base." *Apollo 11* Lunar Science Conference. Proceedings. Vol. 3, Supplement 1, 2399–12.

1971

"Origin of Fragmental Debris on the Lunar Surface and the History of Bombardment of the Moon." Instituto de Investigaciones Geologicas de la Diputacion Provincial. Universidad de Barcelona. Vol. 25, 27–56.

1972

with Stuart-Alexander, D. E., and Moore, H. J. "Geologic Map of the Mule Ear Diatreme, San Juan County, Utah." U.S. Geological Survey Map I-674.

1974

with Kieffer, S. W. *Guidebook to the Geology of Meteor Crater.* Prepared for the Thirty-seventh Annual Meeting of the Meteoritical Society. August 4, p. 66.

1975

and Swann, G. A., Continental drilling, report of the workshop on continental drilling: ed., Carnegie Institution of Washington, p. 55.

1976

with Helin, E. F., and Gillett, S. L. "Populations of Planet-Crossing Asteroids." *Geologica Romana.* Vol. 15, 487–89.

1977

"Astronomically Observable Crater-Forming Projectiles." In Roddy, D. J., Pepin, R. O., and Merrill, R. B., eds. *Impact and Explosion Cratering: Planetary and Terrestrial Implications.* New York: Pergamon Press, 617–28.
"Why Study Impact Craters? In Roddy, D. J., Pepin, R. O, and Merrill, R. B., eds. *Impact and Explosion Cratering: Planetary and Terrestrial Implications.* New York: Pergamon Press, 1–10.
with Helin, E. F. "Populations of Planet-Crossing Asteroids and the Relationship of *Apollo* Objects to Main-Belt Asteroids and Comets. In Delsemme, A. H., ed. *Comets, Asteroids, Meteorites: Interrelations, Evolution, and Origins.* Toledo, Ohio: University of Toledo, 297–300.

1978

"Search for Near-Earth Asteroids." In Arnold, J. R., and Duke, M. B., eds. Summer workshop on near-Earth resources. National Aeronautics and Space Administration Conference Publication 2031, 57–61.
with Helin, E. F. "Earth-Approaching Asteroids: Populations, Origin, and Compositional Types. In Morrison, D., and Wells, W. C., eds. *Asteroids: An Exploration Assessment.* National Aeronautics and Space Administration Conference Publication 2053, 161–76.
with Squires, R. L., and Abrams, M. J. "The Bright Angel and Mesa Butte Fault System of Northern Arizona. In Geology of Northern Arizona. Geological Society of America. Rocky Mountain Society Meeting. Flagstaff, Arizona, 355–91.
with Smith, B. A. "Dynamics of Volcanic Plumes on Io." *Nature.* Vol. 280, 743–46.

with Williams, J. G., Helin, E. F., and Wolfe, R. F. Earth-Crossing Asteroids: Orbital Classes, Collision Rates with Earth and Origin. In Gehrels, T., ed. *Asteroids*. Tucson: University of Arizona Press, 253–82.

with Smith, B. A., Kieffer, S. W., and Cook, A. F., II. "The Role of SO2 in Volcanism on Io." *Nature*. Vol. 280, 738–43.

1980

with Paurcker, M. E., Elston, D. P. "Early Acquisition of Characteristic Magnetization in Red Beds of the Moenkopi Formation (Triassic), Gray Mountain, Arizona." *Journal of Geophysical Research*. Vol. 85, no. N2, 997–1012.

1981

with Emiliani, C., Kraus, E. B. "Sudden Death at the End of the Mesozoic." *Earth and Planetary Science Letters*. Vol. 55, 317–34.

"The Collision of Solid Bodies. In Beatty, J. K., O'Leary, B., and Chaikin, A., eds. *The New Solar System*. Cambridge, Mass.: Sky Publishing Corporation, 33–45.

"Lunar Geology." In Hanle, P., and Chamberlain, V., eds., *Space Science Comes of Age: Perspectives in the History of the Space Sciences*. National Air and Space Museum, Washington, D.C.: Smithsonian Institution, 51–57.

1982

with Passey, Q. R., "Craters and Basins on Ganymede and Callisto: Morphological Indicators of Crustal Evolution. In Morrison, D., ed. *The Satellites of Jupiter*. Tucson: University of Arizona Press, 379–434.

with Lucchitta, B. K., Plescia, J. B., Squyres, S. W., and Wilhelms, D. E. "Geology of Ganymede." In Morrison, D., ed. *The Satellites of Jupiter*, 435–520.

with Wolfe, R. F. "Cratering Time Scales for the Galilean Satellites of Jupiter. In ibid., 277–339.

1983

"Asteroid and Comet Bombardment of the Earth." *Annual Review of Earth and Planetary Sciences*. Vol. 11, 461–94.

1984

"Large Body Impacts through Geologic Time." In Holland, H. D., and Trendall, A. F., eds. *Patterns of Change in Earth Evolution*. Dahlem Konferenzen: Berlin Springer-Verlag, 15–40.

1985

with Carrier, G. F., Moran, W. J., Decker, R. W., Eardley, D. M., Friend, J. P., Jones, E. M., Katz, J. I., Keeny, S. M., Jr., Leovy, C. B., Longmire, C.L., McElroy, M. B., Press, W., Ruina, J. P., Smith, L., Toon, O. B., and Turco, R. P. *The Effects on the Atmosphere of a Major Nuclear Exchange.* Washington, D.C.: National Academy Press, p. 193.

with Shoemaker, C. S. "Recent Discoveries of Comets with the Palomar 46-cm Schmidt Camera." *International Comet Quarterly.* Vol. 7, 3–7; also in Reports of Planetary Geology and Geophysics Program, National Aeronautics and Space Administration Technical Memorandum 87563, 591–93.

1986

with Wolfe, R. F. "Mass Extinctions, Crater Ages, and Comet Showers. In Smoluchowski, R. Bahcall, J. N., and Matthews, M., eds. *The Galaxy and the Solar System.* Tucson: University of Arizona Press, 338–86.

with Tanaka, K. L., Ulrich, G. E., and Wolfe, E. W. "Migration of Volcanism in the San Francisco Volcanic Field, Arizona." *Geological Society of America Bulletin.* Vol. 97, no. 2, 129–41.

1987

with Hut, P., Alvarez, W., Elder, W. P., Hanson, T., Kauffman, E. G., Keller, G., and Weissman, P. R. "Comet Showers As a Cause of Mass Extinctions." *Nature.* Vol. 329, 118–26.

with Schaber, G. G., and Kozak, R. C. "The Surface Age of Venus: Use of the Terrestrial Cratering Record." *Solar System Research.* Vol. 21, 89–94.

with Pillmore, C. L., and Peacock, E. W. "Remanent Magnetization of Rocks of Latest Cretaceous and Earliest Tertiary Age from Drill Core at York Canyon, New Mexico." Geological Society of America Special Paper 190, 131–50.

with Stephens, H. G. *In the Footsteps of John Wesley Powell: An Album of Comparative Photographs of the Green and Colorado Rivers, 1871–72.* Boulder, Colo.: Johnson Books, p. 286.

1988

"A Note about the Illustrations." In Cooley, John. *The Great Unknown: The Journals of the Historic First Expedition Down the Colorado River.* Flagstaff, Ariz.: Northland Publishing, ix–x.

1989

with Alvarez, W., Hansen, T., Hut, P., and Kauffman, E. G. "Uniformitarianism and the Response of Earth Scientists to the Theory of Impact Crises. In Clube, S. V. M., ed. *Catastrophes and Evolution: Astronomical Foundations.* Cambridge, England: Cambridge University Press, 13–24.

with Shoemaker, C. S., and Wolfe, R. F. "Trojan Ssteroids: Populations, Dynamical Structure and Origin of the L4 and L5 Swarms. In Binzel, R. P., and Matthews, M.S., eds. *Asteroids II.* Tucson, Ariz.: University of Arizona Press, 487–523.

with Wolfe, R. F., and Shoemaker, C. S. "Asteroid and Comet Flux in the Neighborhood of the Earth. In Sharpton, V. L., and Ward, P. D., eds. Global Catastrophes in Earth History: An Interdisciplinary Conference on Impacts, Volcanism, and Mass Mortality. Geological Society of America Special Paper 247, 155–70.

1991

with Nishizumi, K., Kohl, C. P., Arnold, J. R., Klein, J., Fink, D., and Middleton, R. "In situ ^{10}Be-^{26}Al Exposure Ages at Meteor Crater, Arizona." *Geochimica et Cosmochimica Acta.* Vol. 55, 2699–703.

1993

with Steiner, M. B., Morales, M. "Magnetostratigraphic, Biostratigraphic, and Lithologic Correlations in Triassic Strata of the Western United States: Applications of Paleomagnetism to Sedimentary Geology." SEPM Spec. Pub. No. 49, 41–57.

1994

Grieve, Richard A. F., and Shoemaker, Eugene M. "The Record of Past Impacts on Earth. In Gehrels, T., ed. *Hazards Due to Comets and Asteroids.* Tuscon: The University of Arizona Press, 417–62.

Rabinowitz, David; Bowell, Edward; Shoemaker, Eugene; and Muinonen, Karri. "The Population of Earth-Crossing Asteroids." In Gehrels, T., ed. *Hazards Due to Comets and Asteroids.* Tuscon: University of Arizona Press, 285–312.

with Plescia, J. B., and Shoemaker, C. S. "Gravity Survey of the Mount Toondina Impact Structure, South Australia." *Journal of Geophysical Research.* Vol. 99, 13, 167–13, 179.

with Robinson, M. S., and Eliason, E. M. "The South Pole Region of the Moon As Seen by *Clementine. Science.* December 16. Vol. 266, 1851–54.

with Shoemaker, C. S., and Levy, David H. "Discovering Comet Shoemaker-Levy 9." In *Once in a Thousand Lifetimes: A Guide to the Collision of Comet Shoemaker-Levy 9 with Jupiter.* Pasadena, Calif.: The Planetary Society, 2–3.

with Weissman, P. R., and Shoemaker, C. S. "The Flux of Periodic Comets Near Earth." In Gehrels, T., ed. *Hazards Due to Comets and Asteroids*, Tuscon: University of Arizona Press, 313–35.

1995

with Levy, D. H., and Shoemaker, C. S. "Comet Shoemaker-Levy 9 Meets Jupiter." *Scientific American.* Vol. 273, 69–75.

"Comet Shoemaker-Levy 9 at Jupiter." *Geophysical Research Letters.* Vol. 22, no. 12, 1555–56;

with Shoemaker, C. S. Foreword for *The Great Comet Crash: The Impact of Comet Shoemaker-Levy 9 on Jupiter,* Spencer, John R. and Mitton, Jacqueline, eds. Cambridge University Press, vii–ix.

with Hassig, P. J., and Roddy, D. J. "Numerical Simulations of the Shoemaker-Levy 9 Impact Plumes and Clouds: A Progress Report:" *Geophysical Research Letters.* Vol. 22, 1825–28.

with Boyarchuk, A. A., Canavan, G., Coradini, M., Darrah, J., Harris, A. J., Morrison, D., Mumma, M. J., Rabinowitz, D. L., Rikhova, R., Chapman, C. R., Marsden, B. G., Ostro, S. J., Worden, S. P., Yeomans, D. K. "Report of the Near-Earth Object Survey Working Group." NASA Solar System Exploration Division, p. 57.

1996

Rabinowitz, David; Bowell, Edward; Shoemaker, Eugene; and Muinonen, Karri. 1994, "The Population of Earth-Crossing Asteroids." In Gehrels, T., ed. *Hazards Due to Comets and Asteroids.* Tuscon: University of Arizona Press, 285–312.

with Shoemaker, C. S. "The Proterozoic Impact Record of Australia." *AGSO Journal of Australian Geology and Geophysics.* Vol. 16, no. 4, 379–98.

1997

with McEwen, A. S., and Moore, J. M. "The Phanerozoic Impact Cratering Rate: Evidence from the Far Side of the Moon." *Journal of Geophysical Research.* Vol. 102, 9231–42.

with Levison, H. F., and Shoemaker, C. S. "Dynamical Evolution of Jupiter's Trojan Asteroids." *Nature.* Vol. 385, 42–44.

with Wynn, J. C. "Secrets of the Wabar Craters." *Sky and Telescope*. Vol. 94, no. 5, 44–48. (includes "Farewell, Gene." Levy, D. H., 48–49).

with Kriens, B. J., and Herkenhoff, K. E. "Structure and Kinematics of a Complex Impact Crater, Upheaval Dome, Southeast Utah." *Brigham Young University Geologic Studies* 42, Part 2. GSA Guidebook, 19.

1998

"Impact Cratering through Geologic Time." *Journal of the Royal Astronomical Society of Canada*. Vol. 92, 297–309.

with Wynn, J. C. "The Day the Sands Caught Fire." *Scientific American*. November, 65–71, 97.

with Farley, K. A., Montanari, A., and Shoemaker, C. S. "Geochemical Evidence for a Comet Shower in the Late Eocene." *Science*. May 22. Vol. 280, 1250–53.

Shoemaker, E. M. and Shoemaker, C. S. "The Role of Collisions." In Beatty, J. K., Petersen, C. C., and Chaikin, Andrew, eds. *The New Solar System*. Cambridge, Mass.: Sky Publishing Corp., 69–85.

Shoemaker, E. M. and Uhlherr, H. Ralph. "Stratigraphic Relations of Australites in the Port Campbell Embayment, Victoria." *Meteoritics and Planet Science*. Vol. 34, 369–84.

with Kriens, B. J., and Herkenhoff, K. E. "Geology of the Upheaval Dome Impact Structure, Southeast Utah." *Journal of Geophysical Research*. Vol. 104, pp. 18867–18887.

CAROLYN SHOEMAKER

1996

"Twelve Years on the Palomar Eighteen-inch Schmidt." *Journal of the Royal Astronomical Society of Canada*. Vol. 90, no. 1, 18–41.

1998

"Space: Where, Now, and Why?" *Science*. Vol. 282, November 27, 1637–38.

1999

"Ups and Downs in Planetary Science." *Annual Review of Earth and Planetary Sciences*. Vol. 27, 1–17.